Mathias Guder

DUV-resist stripping with ozonated water in semiconductor industry

Mathias Guder

DUV-resist stripping with ozonated water in semiconductor industry

Process chemistry and analytical aspects

Südwestdeutscher Verlag für Hochschulschriften

Impressum/Imprint (nur für Deutschland/ only for Germany)
Bibliografische Information der Deutschen Nationalbibliothek: Die Deutsche Nationalbibliothek verzeichnet diese Publikation in der Deutschen Nationalbibliografie; detaillierte bibliografische Daten sind im Internet über http://dnb.d-nb.de abrufbar.

Alle in diesem Buch genannten Marken und Produktnamen unterliegen warenzeichen-, marken- oder patentrechtlichem Schutz bzw. sind Warenzeichen oder eingetragene Warenzeichen der jeweiligen Inhaber. Die Wiedergabe von Marken, Produktnamen, Gebrauchsnamen, Handelsnamen, Warenbezeichnungen u.s.w. in diesem Werk berechtigt auch ohne besondere Kennzeichnung nicht zu der Annahme, dass solche Namen im Sinne der Warenzeichen- und Markenschutzgesetzgebung als frei zu betrachten wären und daher von jedermann benutzt werden dürften.

Verlag: Südwestdeutscher Verlag für Hochschulschriften Aktiengesellschaft & Co. KG
Dudweiler Landstr. 99, 66123 Saarbrücken, Deutschland
Telefon +49 681 37 20 271-1, Telefax +49 681 37 20 271-0
Email: info@svh-verlag.de
Zugl.: Frankfurt/Main, Goethe-Universität, Diss., 2010

Herstellung in Deutschland:
Schaltungsdienst Lange o.H.G., Berlin
Books on Demand GmbH, Norderstedt
Reha GmbH, Saarbrücken
Amazon Distribution GmbH, Leipzig
ISBN: 978-3-8381-1500-9

Imprint (only for USA, GB)
Bibliographic information published by the Deutsche Nationalbibliothek: The Deutsche Nationalbibliothek lists this publication in the Deutsche Nationalbibliografie; detailed bibliographic data are available in the Internet at http://dnb.d-nb.de.

Any brand names and product names mentioned in this book are subject to trademark, brand or patent protection and are trademarks or registered trademarks of their respective holders. The use of brand names, product names, common names, trade names, product descriptions etc. even without a particular marking in this works is in no way to be construed to mean that such names may be regarded as unrestricted in respect of trademark and brand protection legislation and could thus be used by anyone.

Publisher: Südwestdeutscher Verlag für Hochschulschriften Aktiengesellschaft & Co. KG
Dudweiler Landstr. 99, 66123 Saarbrücken, Germany
Phone +49 681 37 20 271-1, Fax +49 681 37 20 271-0
Email: info@svh-verlag.de

Printed in the U.S.A.
Printed in the U.K. by (see last page)
ISBN: 978-3-8381-1500-9

Copyright © 2010 by the author and Südwestdeutscher Verlag für Hochschulschriften Aktiengesellschaft & Co. KG and licensors
All rights reserved. Saarbrücken 2010

Acknowledgements

As is often the case, a task undertaken by one person is not completed successfully without the help and support of numerous persons along the way. To all those who have helped in one way or another to make this dissertation possible I owe a debt of gratitude.

Most of all I would like to thank Prof. Dr. B. O. Kolbesen without whom this project would not have been launched and who saw me through the organizational part of it. I am deeply grateful to him for his valuable scientific advice and guidance and for enabling me to present my work at various conferences and meetings and thus meet people of similar interests.

I am equally grateful to SEZ AG Austria, our industrial partner in this project, for the financial and technical assistance received. I would like in particular to thank the project manager from SEZ AG, Dr. Michael Dalmer and Mrs. Sally-Anne Henry, an expert in the field of resist technology for their advice and support.

I am also indebted to "IMEC", and Rita Vos in particular, for preparing and providing me with implanted resist wafers.

I would like to thank all the members of the research/study group of Prof. Kolbesen for their cooperation, companionship and good humour. My special thanks go to Yvonne Filbrandt-Rozario for her infinite patience during the correction and refinement of the English of my manuscripts, including this dissertation.

I would further like to thank my trainees Maren Pellowska, Maximilian Pohland, Florian Schäfer, Matthias Berger and Julia Herz, who often had the unrewarding task of repeating experiments for statistical coverage.

Last but not least, I would like to thank my parents for their unconditional support and encouragement during the entire course of my studies. Without them things would have been a lot more difficult.

To sum up the efforts of all involved in this project I would like to quote President Obama "Yes we can!"

Danksagung

Wie in den meisten Fällen wird keine Arbeit nur von einer Person erledigt, sondern es gibt immer Leute im Hintergrund die das eine oder andere zum Gelingen beigetragen haben. Warum sollte es hier also anders sein? Daher gebührt auch hier einigen Leuten mein Dank.

Allen voran danke ich meinem Doktorvater Herrn Prof. Dr. B. O. Kolbesen ohne den dieses Projekt gar nicht erst zu Stande gekommen wäre und der mir einige Male, besonders bei organisatorischen Fragen sehr geholfen hat. Aber auch sein Wissen und seine Diskussionsbereitschaft haben mich ein ums andere Mal weiter gebracht. Nicht zu Letzt verdanke ich ihm die Möglichkeit meine Ergebnisse auf Konferenzen und Treffen vorstellen zu können, um so einen ersten Schritt in die Wissenschaftsgemeinde zu tun und mit seiner Hilfe neue Kontakte knüpfen zu können.

Als zweites, aber nichts desto weniger, gilt mein Dank meinem Kooperationspartner, der Firma „SEZ AG Austria" sowohl für die Finanzierung meiner Arbeit als auch die Bereitstellung aller benötigten Geräte sowie der Proben. Mein Dank gilt dabei besonders Herrn Dr. Michael Dalmer als Projektleiter seitens SEZ und Frau Sally-Ann Henry als Expertin auf dem Gebiet der „Resist"-Technologie.

Als weiterem Partner möchte ich IMEC und dabei besonders Frau Dr. Rita Vos für die Präparation und Zurverfügungstellung der implantierten Lack-Proben danken.

Neben meinem Doktorvater und meinen Kooperationspartnern möchte ich selbstverständlich auch meinen Kollegen aus dem Arbeitskreis Kolbesen für die angenehme Zeit während meiner Doktorarbeit danken. Besonders danke ich Frau Yvonne Filbrandt-Rozario für ihre Mühen und ihre Geduld bei der Korrektur und sprachlichen Verbesserung meiner englischen Manuskripte und besonders dieser Arbeit.

Neben meinen Kollegen danke ich auch meinen Praktikanten Maren Pellowska, Maximilian Pohland, Florian Schäfer, Matthias Berger und Julia Herz, denen oft die undankbare Aufgabe zugefallen ist einige Experimente statistisch abzusichern.

Zum Abschluss möchte ich meinen Eltern für Ihre Unterstützung über die gesamte Zeit des Studiums danken, ohne deren Hilfe vieles schwieriger geworden wäre.

Als Fazit für alle Beteiligten an diesem Projekt möchte ich ziehen: „Yes we can."

Table of content

Abbreviations 11
1. Introduction and motivation 13
2. Theory 14
   2.1. Photo resists & lithography 14
   2.2. Ozone - properties and behaviour 26
   2.3. Radical determination 37
      2.3.1. The DDL-method 37
         2.3.1.1. Trapping with MeOH 38
         2.3.1.2. Trapping with DMSO 40
      2.3.2. The combined approach of iodometry and UV/Vis spectroscopy 41
   2.4. IR spectroscopy 43
   2.5. Raman microscopy 47
3. Experimetal Details 50
   3.1. UV/Vis spectroscopic determination of ozone decomposition 55
   3.2. Radical determination 59
      3.2.1. The DDL-method 59
         3.2.1.1. Trapping with MeOH 61
         3.2.1.2. Trapping with DMSO 62
      3.2.2. The combined approach of iodometry and UV/Vis spectroscopy 63
   3.3. Resist characterization with IR spectroscopy 64
   3.4. Resist characterization with Raman microscopy 65
   3.5. Resist stripping 66
4. Results and discussion 69
   4.1. UV/Vis spectroscopic determination of ozone decomposition 69
   4.2. Radical determination 82
      4.2.1. The DDL-method 82
         4.2.1.1. Trapping with MeOH 83

Table of content

        4.2.1.2.   Trapping with DMSO 87

    4.2.2.   The combined approach of iodometry and UV/Vis spectroscopy 92

  4.3.   Resist characterization with IR spectroscopy 94

  4.5.   Resist characterization with Raman microscopy 103

  4.6.   Resist stripping 108

5.   Summary / Zusammenfassung 127

6.   Outlook 135

7.   Literature references 137

8.   Chemicals and equipment 140

9.   Appendix 143

  9.1.   Detailed theory for negative resists 143

  9.2.   UV/Vis spectroscopic determination of ozone decomposition 149

  9.3.   Resist characterization with IR spectroscopy 156

  9.5.   Resist stripping 159

List of figures and tables

Figure 1 - Lithographic process flow [1]...14

Figure 2 - Positive & negative resist technique [1]..15

Figure 3 - Dissolution promotion mechanism for DNQ type positive resists.................17

Figure 4 - Silyl protective groups for OH functions [4]..18

Figure 5 - Ester and carbonate protective groups for OH functions [4].......................19

Figure 6 - Photo acid generators (PAG's) 20

Figure 7 - t-Boc deprotection mechanism (t-Boc → CO_2)......................................21

Figure 8 - Roadmap for photolithography [1]..23

Figure 9 - Wavelengths from a mecury lamp (g-, h- and i-line) [1].............................23

Figure 10 - Resist structures (i-line vs. DUV)..24

Figure 11 - Ozone reactivity in water towards functional organic groups [7]...............25

Figure 12 - MO orbitals of O_2 [12]..28

Figure 13 - Electronic and molecular structure of O_3 [8].......................................29

Figure 14 - UV/Vis-spectrum of ozone...30

Figure 15 - Ozonolysis reaction mechanisms 31

Figure 16 - Mechanism of O_3 decomposition at pH < 4.......................................32

Figure 17 - Mechanism of O_3 decomposition at pH 4 - 8.....................................33

Figure 18 - Mechanism of O_3 decomposition at pH > 10.....................................34

Figure 19 - Mechanism of radical addition...35

Figure 20 - Mechanism of radical substitution 35

Figure 21 - Mechanism of hydrogen abstraction 36

Figure 22 - Mechanism for DDL formation...37

Figure 23 - H_2O_2 decomposition by UV light..38

Figure 24 - Mechanism of radical trapping with MeOH 38

Figure 25 - AOP mechanism in acidic and neutral media 39

Figure 26 - AOP mechanism in alkaline media 40

List of figures and tables

Figure 27 - Mechanism of radical trapping with DMSO 40

Figure 28 - Schematic for combined iodometric titration and UV/Vis approach 41

Figure 29 - IR spectra for Si 43

Figure 30 - IR transmission setup 44

Figure 31 - directed reflection-IR setup 44

Figure 32 - ATR-IR setup 45

Figure 33 - Scheme for crusted resist 47

Figure 34 - Raman spectrum of a Si-wafer 47

Figure 35 - Raman spectra: gem quality diamond(top), glassy carbon(bottom) 48

Figure 36 - Raman spectra: highly ordered pyrolitic graphite(top), conventional graphite(bottom) 49

Figure 37 - PHS-PG 51

Figure 38 - PHS-tBoc 51

Figure 39 - Ethyl lactate 51

Figure 40 - PMA 51

Figure 41 - Ozone module from SEZ 53

Figure 42 - Sampling setup 53

Figure 43 - General setup of the water bath 53

Figure 44 - UV/Vis-measurement setup 53

Figure 45 - Temperature of solution in beaker at given water bath temperatures 54

Figure 46 - Sample holder for resist stripping in beaker 66

Figure 47 - Setup for dispensed DI/O_3 application with additives and UV radiation 67

Figure 48 - Hägg-diagram for H_2CO_3 71

Figure 49 - [O_3] vs. t; RT; recirculation; different pH values 79

Figure 50 - [O_3] vs. t; 50 °C; recirculation; different pH values 79

Figure 51 - ln [O_3] vs. t; RT; recirculation; different pH values 80

Figure 52 - ln [O_3] vs. t; 50 °C; recirculation; different pH values 80

Figure 53 - 1/[O_3] vs. t; RT; recirculation; different pH values 81

Figure 54 - 1/[O_3] vs. t; RT; recirculation; different pH values 81

List of figures and tables

Figure 55 - DDL development and stability 82

Figure 56 - Calibration of CH_2O conversion to DDL with MeOH in excess 83

Figure 57 - Calibration of radical trapping with MeOH in diluted & concentrated form 84

Figure 58 - Calibration of radical trapping with MeOH with standard addition 85

Figure 59 - Radical determination at different pH with MeOH 86

Figure 60 - Calibration of CH_2O conversion to DDL with DMSO in excess 87

Figure 61 - Calibration of radical trapping with DMSO in diluted & concentrated form 88

Figure 62 - Calibration of radical trapping with DMSO with standard addition 89

Figure 63 - Radical determination at different pH with DMSO 90

Figure 64 - Comparison of IR spectra of M91Y; UV26; DUV248 as applied 94

Figure 65 - Resist changes during exposure 95

Figure 66 - Resist changes during PEB 96

Figure 67 - Resist changes during plasma etch 97

Figure 68 - IR spectrum of JSR KrF M91Y before and after ozone treatment 98

Figure 69 - IR spectrum of Rohm&Haas UV26 before and after ozone treatment 99

Figure 70 - Decomposition of the Rohm&Haas UV26 resist 100

Figure 71 - IR spectrum of JSR DUV248 5 keV before and after ozone treatment 101

Figure 72 - IR spectrum of JSR DUV248 40 keV before and after ozone treatment 102

Figure 73 - Raman overview spectra of DUV248 5 and 40 keV 103

Figure 74 - Raman spetrum of DUV248 As 10^{15} 5 keV vs. blank Si 104

Figure 75 - Raman spetrum of DUV248 As 10^{16} 40 keV vs. blank Si 104

Figure 76 - Raman spectra of DUV248 5 keV with different depths of focus 105

Figure 77 - Raman spectra of DUV248 40 keV with different depths of focus 105

Figure 78 - DUV248 As 10^{15} 5 keV peak allocation 106

Figure 79 - DUV248 As 10^{16} 40 keV peak allocation 106

Figure 80 - pH effect on stripping efficiency of DI/O_3 on UV26 at 25 °C/50 ° (by pH) 108

Figure 81 - pH effect on stripping efficiency of DI/O_3 on UV26 at 25 °C/50 ° (by pH) 109

Figure 82 - pH effect on stripping efficiency of DI/O_3 on UV26 at 25 °C/50 ° (by step) 110

List of figures and tables

Figure 83 - pH effect on stripping efficiency of DI/O$_3$ on UV26 at 25 °C/50 ° (by step) 110

Figure 84 - Phenol deprotonation 112

Figure 85 - pH effect on stripping efficiency of DI/O$_3$ on M91Y/UV26 at 25 °C (by pH) 113

Figure 86 - pH effect on stripping efficiency of DI/O$_3$ on M91Y/UV26 at 25 °C (by pH) 113

Figure 87 - pH effect on stripping efficiency of DI/O$_3$ on M91Y/UV26 at 25 °C (by step) 114

Figure 88 - pH effect on stripping efficiency of DI/O$_3$ on M91Y/UV26 at 25 °C (by step) 114

Figure 89 - UV26 resist thinning by DI/O$_3$ at 25 °C 115

Figure 90 - Surface scan 10000 µm 116

Figure 91 - Surface scan 1000 µm 116

Figure 92 - UV26 DI/O$_3$ for 5 min 117

Figure 93 - UV26 DI/O$_3$ for 10 min 117

Figure 94 - (As 10^{15} cm^{-2}; 5 keV) – DI/O$_3$ pH=3.5; beaker; 90 °C; 1 h 118

Figure 95 - (As 10^{15} cm^{-2}; 5 keV) – DI/O$_3$ + HAc-buffer pH=5; beaker; 90 °C; 1 h 119

Figure 96 - (As 10^{15} cm^{-2}; 5 keV) – DI/O$_3$ + PO$_4^{3-}$-buffer pH=6; beaker; 90 °C; 1 h 119

Figure 97 - (As 10^{15} cm^{-2}; 5 keV) – DI/O$_3$ flow + UV; 10 min 120

Figure 98 - (As 10^{15} cm^{-2}; 5 keV) – DI/O$_3$ flow; 10 min 121

Figure 99 - (As 10^{15} cm^{-2}; 5 keV) – DI flow + UV; 10 min 121

Figure 100 - (As 10^{16} cm^{-2}; 40 keV) – DI/O$_3$ flow + UV; 1 h 122

Figure 101 - (As 10^{15} cm^{-2}; 5 keV) – KOH pH=13.5 flow; 3 min 123

Figure 102 - (As 10^{16} cm^{-2}; 40 keV) – KOH pH=13.5 flow; 1 h 124

Figure 103 - (As 10^{16} cm^{-2}; 40 keV) – Pyrrolidine pH=13.12 flow; 45 min 125

Figure 104 - (As 10^{16} cm^{-2}; 40 keV) - hard bake 130 °C for 10 s 126

Figure 105 - (As 10^{16} cm^{-2}; 40 keV) – hard bake, DI/O$_3$ + NH$_4$OH (pH=13) flow + UV 126

Figure 106 - Photo acid generators (PAG's) 143

Figure 107 - Crosslinkers for AHR resists 144

Figure 108 - Mechanism of acid hardening in resists 145

Figure 109 - General mechanism of photopolymerization 146

Figure 110 - Norrish type I radical starters 147

List of figures and tables

Figure 111 - Norrish type II radical starters..................147

Figure 112 - Lewis acid induced cationic ring opening polymerzation..................148

Figure 113 - $[O_3]$ vs. t; RT; mixing effects; pH=1; H_2SO_4..................149

Figure 114 - $[O_3]$ vs. t; 50 °C; mixing effects; pH=1; H_2SO_4..................149

Figure 115 - $[O_3]$ vs. t; RT; mixing effects; pH=1; H_3PO_4..................150

Figure 116 - $[O_3]$ vs. t; 50 °C; mixing effects; pH=1; H_3PO_4..................150

Figure 117 - $[O_3]$ vs. t; RT; mixing effects; pH=2.3; HF..................151

Figure 118 - $[O_3]$ vs. t; 50 °C; mixing effects; pH=2.3; HF..................151

Figure 119 - $[O_3]$ vs. t; RT; mixing effects; pH=3.5; pure..................152

Figure 120 - $[O_3]$ vs. t; 50 °C; mixing effects; pH=1; pure..................152

Figure 121 - $[O_3]$ vs. t; RT; mixing effects; pH=3.6; H_2CO_3..................153

Figure 122 - $[O_3]$ vs. t; 50 °C; mixing effects; pH=3.6; H_2CO_3..................153

Figure 123 - $[O_3]$ vs. t; RT; mixing effects; pH=7.6; HCO_3^-..................154

Figure 124 - $[O_3]$ vs. t; RT; mixing effects; pH=9; NH_4OH..................155

Figure 125 - Comparison of IR spectra M91Y; UV26 after exposure..................156

Figure 126 - Comparison of IR spectra M91Y; UV26 after PEB..................157

Figure 127 - Comparison of IR spectra M91Y; UV26 after plasma..................158

Figure 128 - Stripping efficiency comparison M91Y 25 °C vs. 50 ° (by pH)..................159

Figure 129 - Stripping efficiency comparison M91Y 25 °C vs. 50 ° (by step)..................159

Figure 130 - Stripping efficiency comparison M91Y vs. UV26 at 50 ° (by pH)..................160

Figure 131 - Stripping efficiency comparison M91Y vs. UV26 at 50 ° (by step)..................160

Figure 132 - (As 10^{16} cm^{-2}; 40 keV) - DI/O_3+buffer pH=5; RT; 1 h; UV; continues DI/O_3 flow...161

Figure 133 - (As 10^{16} cm^{-2}; 40 keV) - DI/O_3+buffer pH=6; RT; 1 h; UV; continues DI/O_3 flow...161

List of figures and tables

Table I	Optical paramaters for lithography	22
Table II	Physical properties (O_2, O_3, H_2O) [9],[10],[11]	27
Table III	Redox equations for O_3 and ·OH reaction with $S_2O_3^{2-}$ [11]	42
Table IV	$S_2O_3^{2-}$ consumption in iodoemtric titration	42
Table V	IR peaks for expected functional groups in the resists[24],[25]	46
Table VI	Wafer preparation steps	52
Table VII	Additives for ozone decomposition studies	56
Table VIII	Dilution series for CH_2O	59
Table IX	Composition of the Hantzsch reagent	60
Table X	[*OH] derived from H_2O_2 decomposition	60
Table XI	Additives for in situ mixing	68
Table XII	Comparison of experimentally obtained constants for k and E_A with hitherto published values	73
Table XIII	Results of ozone decomposition studies	74
Table XIV	*OH conc. derived: MeOH vs. DMSO, standard addition, different irradition times	90
Table XV	*OH conc. derived: MeOH vs. DMSO by standard addition at different pH values	91
Table XVI	Results of radical determination with the iodoemtric-UV/Vis approach	92
Table XVII	Comparison of trapping results for DI/O_3 at pH=1	93
Table XVIII	Oxidising potentials (determined by potentiometric measurements)	111
Table XIX	Figure comparison of 1 h stripping pH=12-13.5 with continues flow	162

Abbreviations

AHR	acid hardening resist
AOP	advanced oxidation process
ARC	anti reflective coating
ATR	attenuated total reflection
Barracuda®	ozonated water vapour wafer cleaning module from „MATTSON WET PRODUCTS"
BEOL	back end of line
DDL	3,5-diacetyl-1,4-dihydrolutidine
DI or DI-water	deionized water
DI/O_3	deionized water with ozone
DMSO	dimethyl sulphoxide
DNQ	2-diazo-1-naphthoquinone
DoF	depth of focus
DPI	diphenyl iodinium
DUV	deep UV
ESR	electron spin resonance
FEOL	front end of line
FT	fourier transform
g-line	436 nm from mercury lamp
h-line	406 nm from mercury lamp
HMDS	hexamethyldisilazane
HMMM	Hexamethoxymethyl melamine
IC	integrated circuit
i-line	365 nm from mercury lamp
IR	infrared specstrocopy
MAH	maleic acid anhydrous
MCPBA	metha-cloroperbenzoic acid
MO orbitals	molecular orbitals
Novolak	phenol based resin
PAC	photo active component
PAG	photo acid generator

Abbreviations

PEB	post exposure bake
PG	protection group / protecting group
PHS	polyhydroxystyrene
PMA	propyleneglycolmethyletheracetate
PMMA	polymethylmethacrylate
RIE	reactive ion etching
RT	room temperature (25 °C)
SET	single electron transfer
SiGe	silicon germanium alloy
SOI	silicon on insulator
SOM	sulphuric acid ozone mixture
SPM	sulphuric acid peroxide mixture
sSOI	strained silicon on insulator
TCD	tetracyclododecene
TCD-alt-MAH	tetracyclododecene alternating maleic acid anhydrous
TPS	triphenyl sulfonium
VOC	volatile organic compounds

1. Introduction and motivation

The ever increasing demand for faster, more energy and cost efficient microelectronic devices and processors which are a crucial part of nearly every modern piece of equipment in industry as well as in the daily life of the private citizen, has been the biggest challenge for the semiconductor industry. To be able to fulfil these demands one of the approaches adopted is the continuous miniaturisation of IC-structures. The other is the introduction of new materials like silicon on insulator (SOI), strained silicon on insulator (sSOI), SiGe for substrates in the front end of line (FEOL), high κ materials for gate dielectrics and low κ materials as insulators for conducting paths in the back end of line (BEOL). Despite their immense advantages these new materials in combination with finer structures generate new problems in nearly every stage of production. This study involves the photolithographic stage in the production and focuses on photoresists which are essential to wafer-structuring. An understanding of the nature of the resists and how they function is a prerequisite for their effective removal, once they have served their purpose.

The photolithographic procedure involves the application and removal of resists in multiple iterations particularly in the BEOL. There are many different resists and classes of resist specially designed for each stage in the process, and the method for their removal varies accordingly. Generally, this was, and in some cases still is, achieved physically by plasma etching with oxygen or fluorine followed by a wet clean supported by ultra or megasonics.

With the dramatic decrease in size of integrated circuit (IC) geometries, with feature sizes reduced to 45 nm in production and 32 nm in research, this method of removing resists is approaching its limitations, being so aggressive for the finer structures as to affect the operational reliability of the product. A non-destructive wet cleaning method for resist removal which would bypass dry pre-treatments such as the plasma etch is desirable. Such alternatives already exist, using sulphuric acid peroxide mixture (SPM), a solution employed in the FEOL to remove organic residues and impurities. Subsequent modifications and developments produced sulphuric acid ozone mixture SOM and later DI/O_3. The suitability of DI/O_3 for resist stripping, its environment-friendliness and cost efficiency make it particularly attractive. Regrettably, modern deep UV (DUV) resists do not respond well to DI/O_3 especially after the ion implantation step. An understanding of the changes that take place in the resist as it goes through the various process steps and trials with additives to DI/O_3 to increase stripping efficiency is the subject of this study.

2. Theory

2.1. Photo resists & lithography

As lithography has proven to be the best way for the transfer of very fine structures on wafer substrates as used in the semiconductor industry it is an essential part of the production of computer chips. Apart from the mechanical components such as the lithography stepper which is a very expensive tool (up to 20 million €), and the mask, which acts as a template, the most important component is the photo resist. Whatever the application, the photo resist is sensitive to the light used for the photolithography. For all resists and applications the principal process flow is as illustrated in Figure 1 [1].

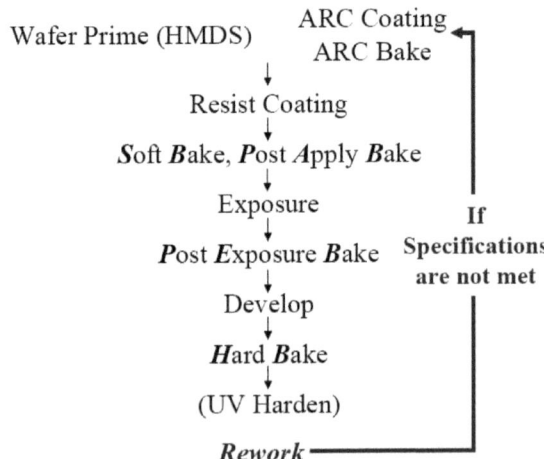

Figure 1 - Lithographic process flow [1]

However there are many differences in the composition and structures of the resists adjusted to the field of application and demands like resolution, thermal stability etc. One basic difference lies in their behaviour towards light and in this sense resists are classified as positive or negative (Figure 2) [1]. A positive resist is one that dissolves in a developer solution exposing the substrate beneath for further processing – whatever shows, goes.

Therefore the mask is designed as an exact copy of the pattern to be transferred to the substrate. The behaviour of the negative resist is the reverse of that of the positive resist, resulting in changing soluble structures into insoluble ones by means of polymerisation during exposure. In this case the mask is a negative of the pattern to be applied. In modern applications bilayer resists are also applied consisting of a sequence of positive and negative resists.

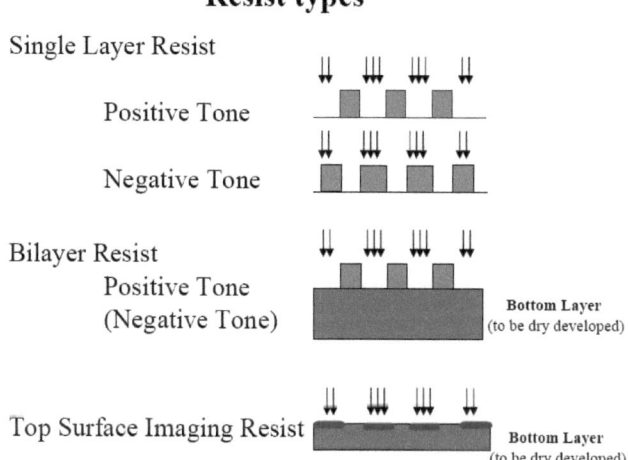

Figure 2 - Positive & negative resist technique [1]

Negative resists [2]:

The theoretical part for the negative resists will be kept short in this chapter as they are not part of this thesis. For more detailed information please see 9.1.

The use of negative resists today is limited to circuits with relatively coarse structures as in earlier times the resolution produced is not up to the mark for today's semiconductor circuits. With ongoing improvements this limitation might be overcome in the future.

Regardless of their structure all negative resists have in common that their solubility decreases with exposure towards light but the manner in which this happens can be very different and therefore they are split into different subgroups of negative resists.

One important group of negative resists are the acid hardening resists (AHRs) as they deliver a high contrast and sensitivity as well as stability towards negative influences like heat and swelling during the processing.

Another way to decrease the solubility in negative resists is to initiate further resist polymerisation through the generation of radicals from radical starters, leading to a classical radical chain reaction pattern (Figure 109).

Beside radicals, cations and anions are also capable of inducing photopolymerization. Their advantages are their insensitivity to oxygen and that cations produce living polymers i.e. there is no chain termination during polymerization by the mechanism itself. They are however very limited in practical application as their disadvantages include sensitivity to termination reactions by nucleophilic impurities, chain-transfer processes and a general lack of initiators.

Positive resists [3]:

As positive resists are poorly soluble to insoluble in the developer before exposure their solubility has to be increased during exposure. In older i-line resists (see i-line vs. DUV) of the Novolak type this is achieved by a two-component system following a dissolution inhibition/promotion mechanism. In this system the second component is transformed from a dissolution inhibitor as applied to a dissolution promoter. Normally 2-diazo-1-naphthoquinone derivatives (DNQs) are used as the dissolution inhibitors (Figure 3). The dissolution inhibition is achieved by the interaction of the unphotolyzed DNQ with the phenolic groups of the Novolak resist. Although this dissolution inhibition is not complete the solubility is quite low and the positive aspects of the Novolak-DNQ-system like high resolution capability, broad processing window and excellent resistance to dry etch processes have made it the first choice for many years.

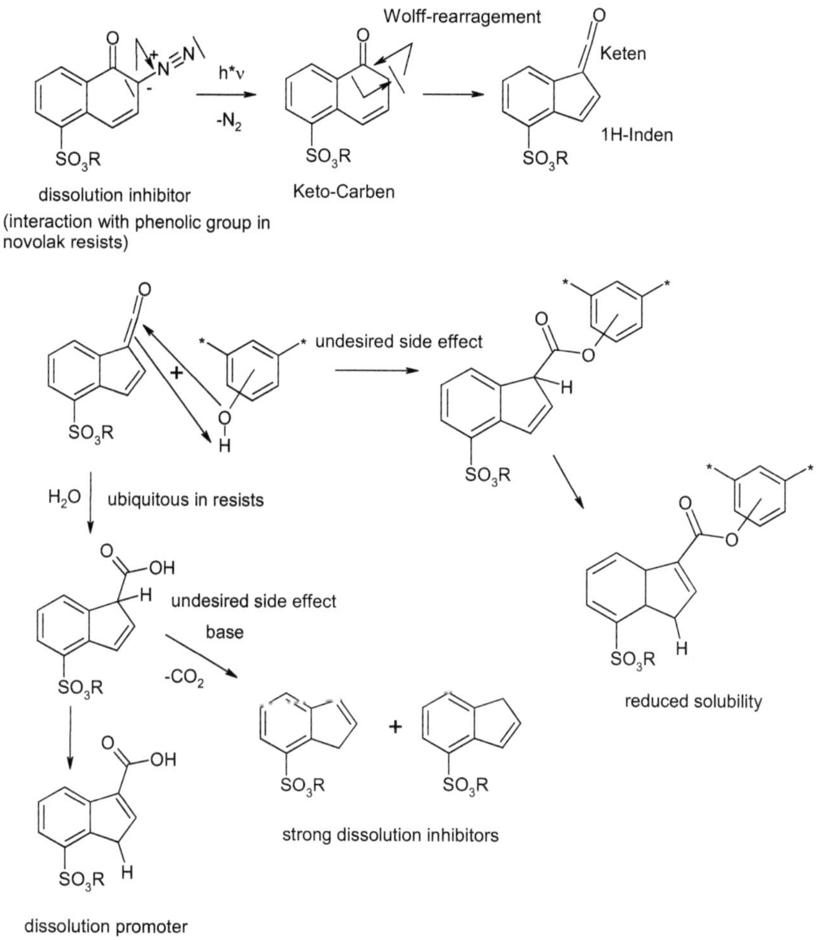

Figure 3 - Dissolution promotion mechanism for DNQ type positive resists

Actually as the resist types have changed from i-line to DUV so have the photo active components (PACs) for the dissolution promotion mechanism. In modern polyhydroxystyrene (PHS) based DUV resists insolubility is achieved by the partial attachment of protective groups (7 -40 %w) to the phenolic OH site. These protective groups display a wide range of structures from silyl ethers (Figure 4) [4] to esters and carbonates (Figure 5) [4].

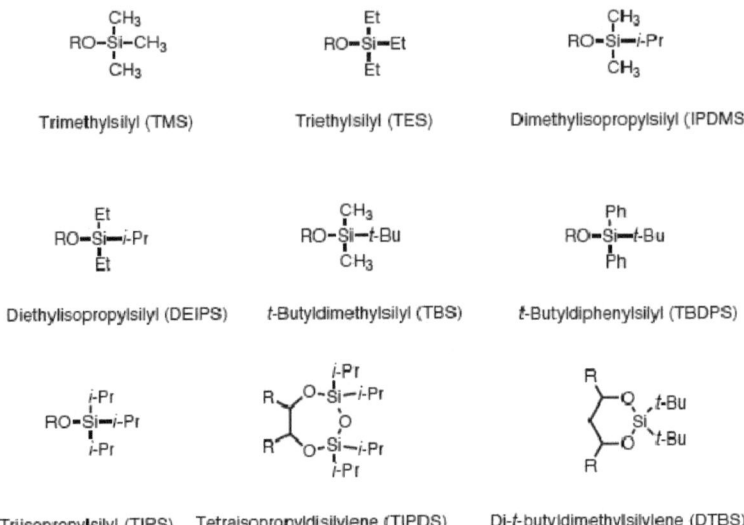

Figure 4 - Silyl protective groups for OH functions [4]

Theory – Photo resists & Photolithography

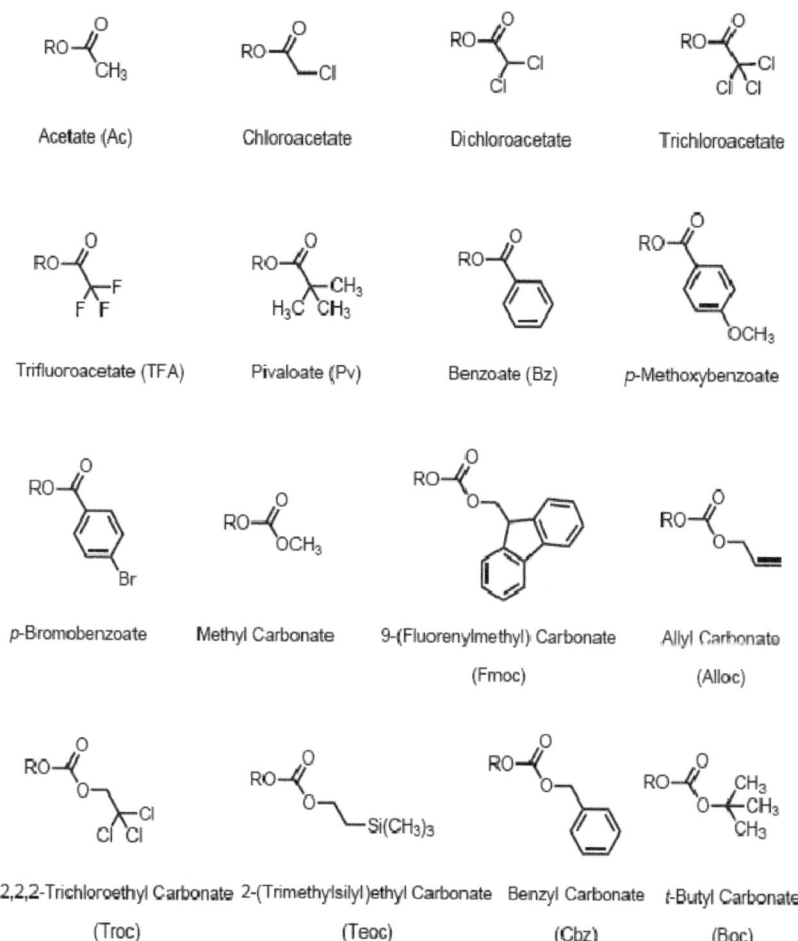

Figure 5 - Ester and carbonate protective groups for OH functions [4]

Theory – Photo resists & Photolithography

The current resists are chosen for their ability to act in the way of acid-catalyzed deblocking. PAGs (Figure 6) (often DPI – diphenyl iodinium or TPS – triphenyl sulfonium) which are used in the negative resists to initiate crosslinking are also employed for acid deblocking but as there are no crosslinkers in positive resists, the acid produced removes the protective groups thereby increasing the solubility of the resist in the developer. The exact mechanism of deblocking depends on the protective group applied and is shown in Figure 7 for the t-BOC/BOC group which is a constituent of the Rohm&Haas UV26 resist [5].

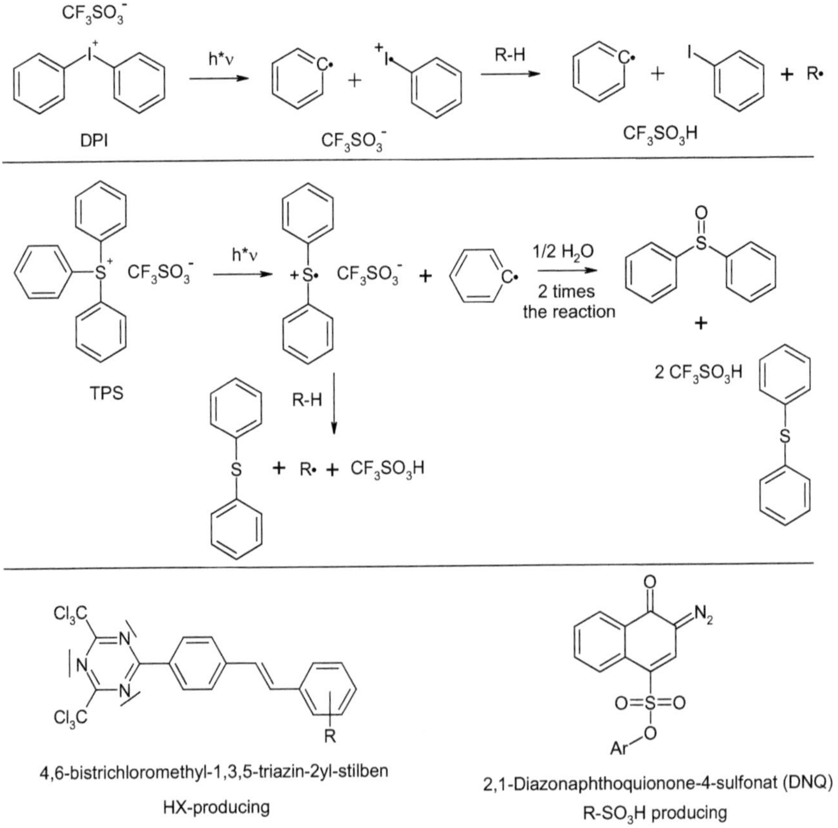

Figure 6 - Photo acid generators (PAG's)

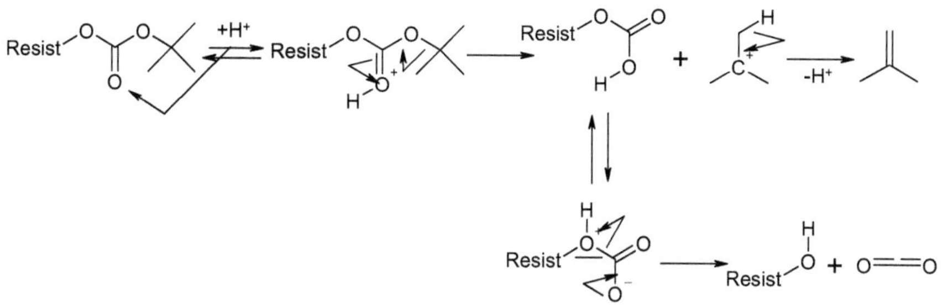

Figure 7 - t-Boc deprotection mechanism (t-Boc → CO_2)

i-line vs. DUV:

Another key issue of the photolithographic process is the wavelength as it not only decides the degree of resolution which translates into the smallest structure that can be produced on a chip as well as the depth of focus (DoF) but especially because it affects the resist structure.

Table I - Optical paramaters for lithography

Resolution	Depth of Focus	Numeric Aperture / Refractive Indices
$\text{Resolution} = k_1 \dfrac{\lambda}{NA}$	$DoF = \dfrac{\lambda}{2\left(1 - \sqrt{1 - NA^2}\right)}$ Depth of Focus	$NA = n \cdot \sin\alpha$ numeric aperture α = angle of aperture
$\text{Resolution} = k_1 \dfrac{\lambda}{n \cdot \sin\alpha}$	$DoF = \dfrac{\lambda/n}{2\left(1 - \sqrt{1 - \dfrac{NA^2}{n^2}}\right)}$	$n_{air} = 1.000292;\ n_{water} = 1.437$ refractive indices

The semiconductor roadmap in Figure 8 [1] shows the development of wavelengths used, beginning with wavelengths of 436 nm (g-line) and decreasing over the years to 406 nm (h-line) and 365 nm (i-line) all derived from the spectra of mercury lamps (Figure 9) [1]. They are currently in the deep UV region (DUV) of 248 nm and 193 nm, the sources of which are a KrF and ArF excimer laser, respectively.

Lithography Roadmap: The Sub-Wavelength Gap

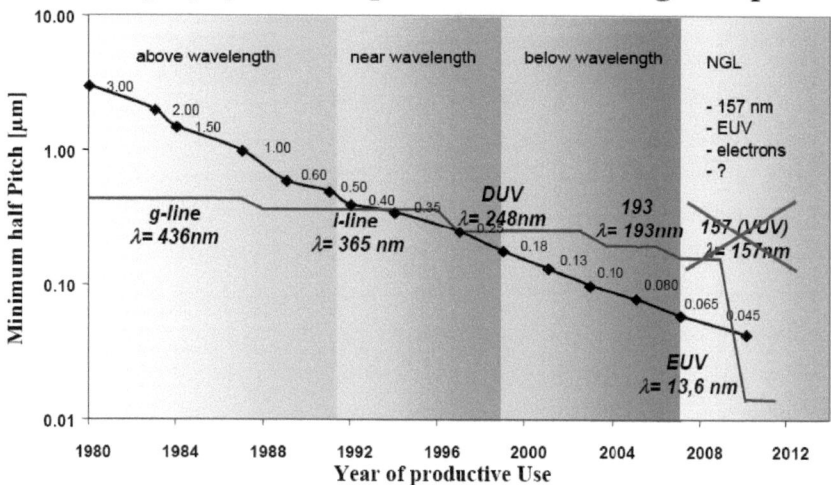

Figure 8 - Roadmap for photolithography [1]

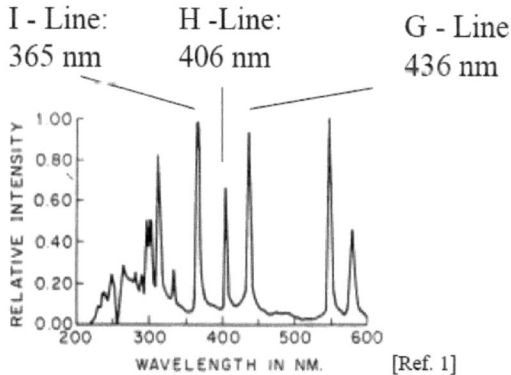

Figure 9 - Wavelengths from a mercury lamp (g-, h- and i-line) [1]

As structural dimensions in current chip technology have reached 45 nm in production we have gone from structures above the wavelength down to structures far below the wavelength. This is only possible with the application of immersion techniques which can vary the gap medium between lens and wafer from air to liquids such as water or even oils all with a refractive index n greater than 1 (Table I). This enables these 45 nm structures to be produced with KrF or ArF light sources.

Theory – Photo resists & Photolithography

This digression to the basics of lithography, its wavelength dependency and changes during structure miniaturisation is necessary as it significantly affects the resists in their chemical composition and structure.

For all the applied resists (regardless if they are positive or negative resists or designed for i-line or DUV) it is essential that their monomeric components – beside the PACs – are transparent towards the light employed for the photolithography as otherwise they would shadow the area below preventing a complete exposure down to the wafer surface. So the change of the light source from mainly i-line to DUV has been accompanied by a change in the resist structure. While i-line resists are mostly of the Novolak type, state-of-the-art KrF-DUV resists are often polyhydroxystyrene (PHS) based and ArF-DUV resists can also be based on poly-tetracyclododecene (poly-TCD) [6] or poly-methylmethacrylate PMMA. And as if these variations wouldn't be enough there are also co-polymers existing as poly-TCD-alt-MAH [6] or combinations of protected and unprotected polymer components as in the Rohm&Haas UV26 resist (examples in Figure 10).

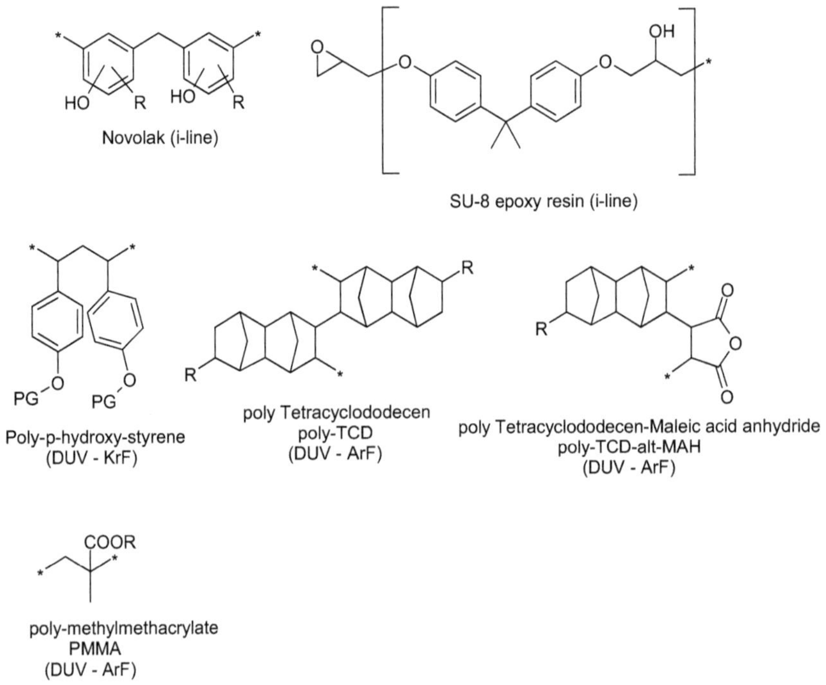

Figure 10 - Resist structures (i-line vs. DUV)

A comparison of all the various structures and alterations will show that the striking change from i-line to DUV resists is the loss of C=C-bonds or their relocation from the polymeric backbone to the side chain. This has a dramatic influence on the stripping efficiency with DI/O_3 as the reactivity of molecular O_3 towards these saturated backbone structures is drastically reduced (Figure 11) [7].

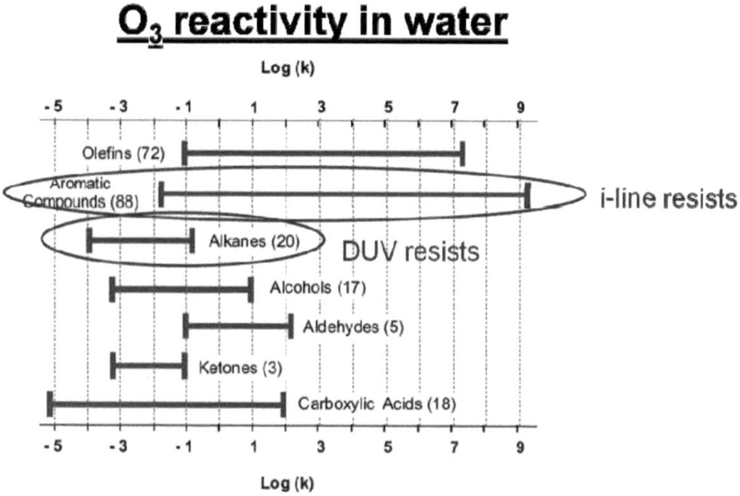

O_3 reactivity in water

Log (k)

```
         -5   -3   -1   1   3   5   7   9
              Olefins (72)  |————————————|
         Aromatic
         Compounds (88) |————————————————————|   i-line resists
              |——————| Alkanes (20)    DUV resists
                     |————————| Alcohols (17)
                            |——————| Aldehydes (5)
                     |——————| Ketones (3)
         |——————————————————| Carboxylic Acids (18)
         -5   -3   -1   1   3   5   7   9
                        Log (k)
```

Figure 11 - Ozone reactivity in water towards functional organic groups [7]

The reason for this huge difference in O_3 reactivity lies in its properties and behaviour.

2.2. Ozone - properties and behaviour

As ozone is the basic component of the stripping solutions to be investigated it is essential to be aware of its properties and reactivity in order to estimate its potential and to plan reasonable experiments.

Ozone derived from the Greek word ozein "to smell" by Shonbein in 1840 was first discovered by Boyle in 1681 when he detected a strong smell during the reaction of air with phosphorus. Later on, in 1785, van Marum observed it coming from electric sparks. The first ozone generator was built by Siemens in 1857 and the chemical structure was established by Soret in 1866. The first known practical application of ozone was in 1893 when the Dutch used it to disinfect drinking water. From this point on ozone has been used for many purposes such as waste water treatment or as a replacement of chlorine in swimming pools. It is also used for special reactions in preparative chemistry. In 1996 ozone found its way into the semiconductor industry having been introduced as chilled ozonized water by the company SMS (Akrion) as a legacy patent. Now it is used by many companies in different variations and processes (for example by Semitool in Boundary Controlled O_3/H_2O in the closed batch spin; by FSI in Mercury and Zeta in the closed batch spin; by Steag (Akrion) in Barracuda® in the closed bath process; by TEL in VOC in the closed bath process; and by NOVO in the OzoneJet open single wafer spin) [8].

The basic components in DI/O_3 are O_2 as the source of O_3, O_3 itself and H_2O as the solvent for the gas. The most striking differences between O_2 and O_3 are their dipole moments, their solubility in water and their standard potential. From the much higher dipole moment of O_3 it is easy to explain the increased solubility of O_3 in the polar solvent H_2O. Bearing in mind that even the non-polar O_2 is soluble in H_2O in an amount that allows fish to breathe the 5 to nearly 10 times higher solubility of O_3 is immense. Another interesting point is the significantly higher standard potential compared to the also oxidative O_2 making O_3 a strong oxidizer.

Theory – Ozone

Table II - Physical properties (O_2, O_3, H_2O) [9], [10], [11]

O_2			
mass $\left[\dfrac{g}{mol}\right]$	Density $\left[\dfrac{g}{m^3}\right]$	melting point [°C]	boiling point [°C]
32	1.43	-218.93	-182.97

solubility $\dfrac{1}{K_H^0}\left[\dfrac{1}{bar*L}\right]$		standard potential [V]	dipole moment [debye]
1.2-1.3x10^{-3}	49.1 $\left[\dfrac{mL}{L}\right]$ 70.2 $\left[\dfrac{mg}{L}\right]$	1.23	0

O_3			
mass $\left[\dfrac{g}{mol}\right]$	Density $\left[\dfrac{kg}{m^3}\right]$	melting point [°C]	boiling point [°C]
48	2.14	-193	-112

solubility $\dfrac{1}{K_H^0}\left[\dfrac{1}{bar*L}\right]$		standard potential [V]	dipole moment [debye]
8.9x10^{-3}-1.3x10^{-2}	200-292 $\left[\dfrac{mL}{L}\right]$ 427-624 $\left[\dfrac{mg}{L}\right]$	E°(O_3+H$^+$/O_2)=2.07V E°(O_3+H$_2$O/O_2)=1.24V	0.53

Theory – Ozone

H₂O			
mass $\left[\dfrac{g}{mol}\right]$	Density $\left[\dfrac{kg}{m^3}\right]$	melting point [°C]	boiling point [°C]
18	1000	0	100

dipole moment [debye]
1.84

In order to understand the properties listed in Table II and to predict possible reactions of O_3 it is necessary to take a closer look at their electronic and molecular structures. The O_2 molecule is linear. The standard triplet O_2 can be described as a diradical according to the MO scheme (Figure 12) [12] and is therefore paramagnetic.

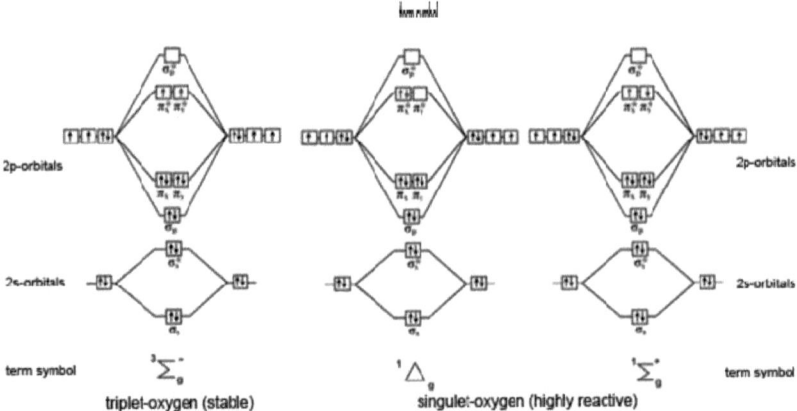

Figure 12 - MO orbitals of O_2 [12]

In contrast the molecular structure of ozone is not linear but has an O-O-O angle of 116.8 °. Its electronic structure is also different. As each oxygen atom has 6 electrons in O_3 18 electrons need to be allocated to the 3 oxygen atoms giving each one its necessary 8 electrons – 18 electrons present, 24 needed. This can only be achieved by the delocalization of π-electrons as shown in Figure 13 [8].

Theory – Ozone

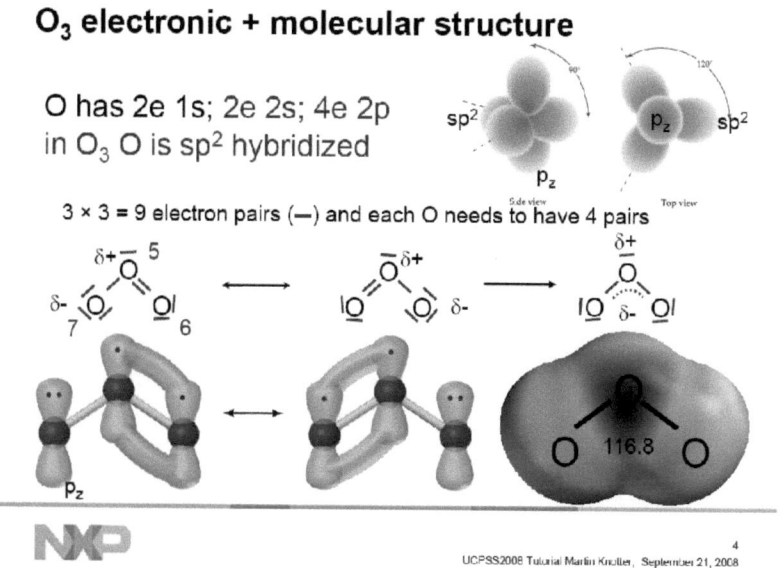

Figure 13 - Electronic and molecular structure of O_3 [8]

This delocalization makes O_3 a 1.3-dipole which is the reason for its classical reactivity.

The delocalized electrons of the O_3 are also responsible for its ability to absorb light. In nature this ability is necessary for life on earth. Ozone in the upper atmosphere acts as a filter for hard and dangerous UV light as it absorbs wavelengths of 256 nm (Hartley band) with a molar absorbance of 3300 $\left[\dfrac{L}{mol*cm}\right]$ [13].

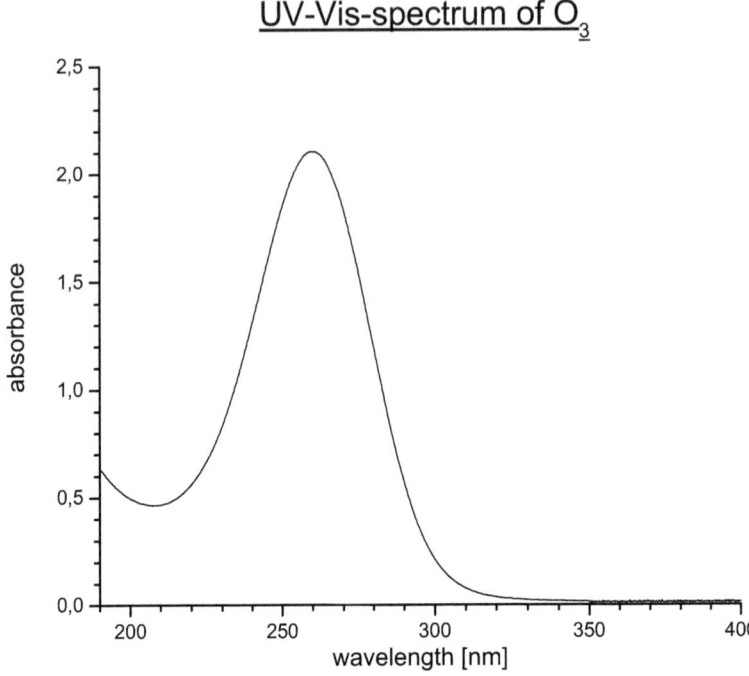

Figure 14 - UV/Vis-spectrum of ozone

Coming back to the reaction pattern of molecular O_3 its reaction is a classic 1.3-cyclo-addition with unsaturated double bonds, mainly C=C-bonds. For O_3 this reaction called ozonolysis [14] results in an oxidative cleavage of the C=C-bonds forming carboxylic acids, ketones, alcohols etc. depending on the follow-up reactions (Figure 15) [14].

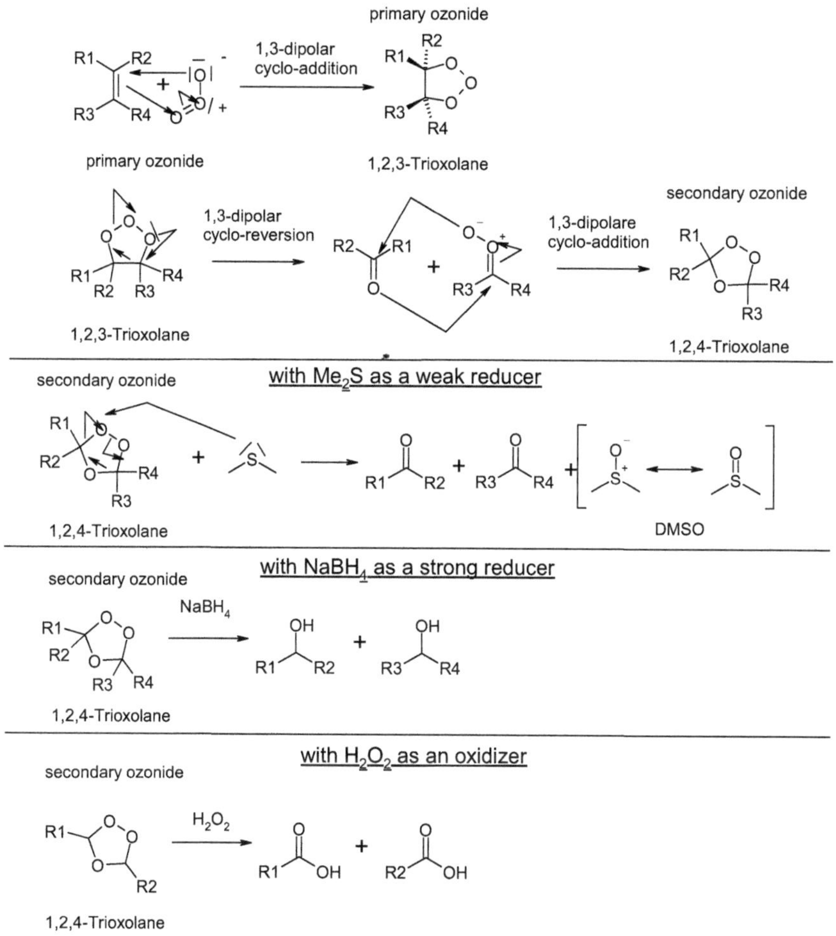

Figure 15 - Ozonolysis reaction mechanisms

Applied to the older i-line resists of the Novolak type with their aromatic C=C-containing phenol groups as part of the polymeric backbone (2.1) this reaction pattern explains the high stripping efficiency of O_3 (Figure 11).

This classical reaction of O_3 molecules occurs mainly in the gas phase. As O_3 is not very stable it decomposes with time to the O_2 from which it originates. In its simplest form this decomposition may be described as the reaction $2O_3 \rightarrow 3O_2$. In the presence of water, however, a more complex reaction takes place, being influenced by several factors particularly the pH value of the solution, and can take the form of three different decomposition pathways that distinguish between 3 different pH regions as proposed by Sehested [15], Virdis [16] and Tomiyasu [17].

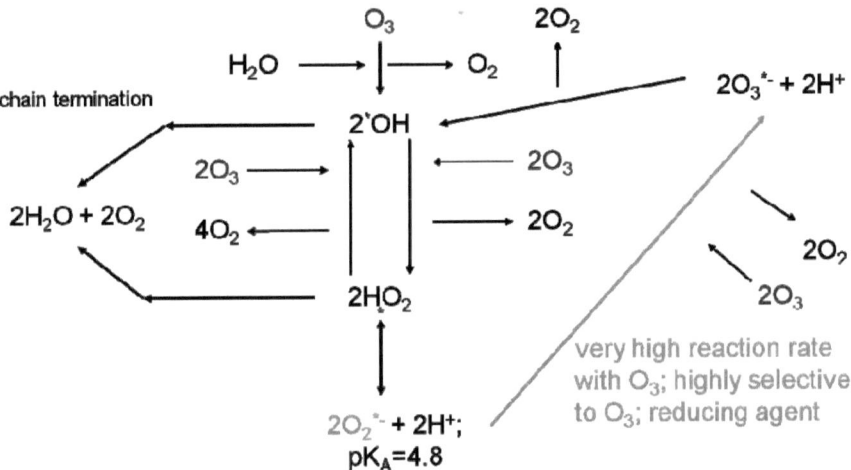

Figure 16 - Mechanism of O_3 decomposition at pH < 4

Theory – Ozone

Decomposition of ozone RHEA-model – pH = 4 – 7.9

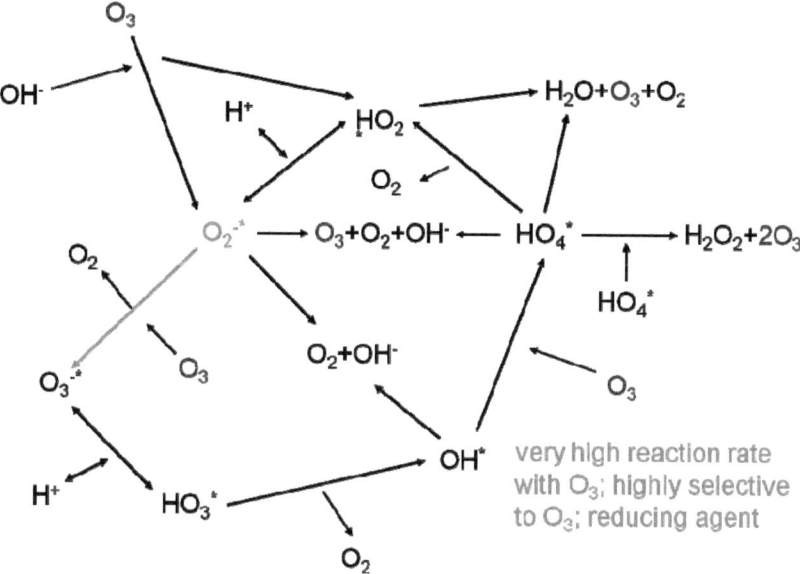

Figure 17 - Mechanism of O_3 decomposition at pH 4 - 8

Theory – Ozone

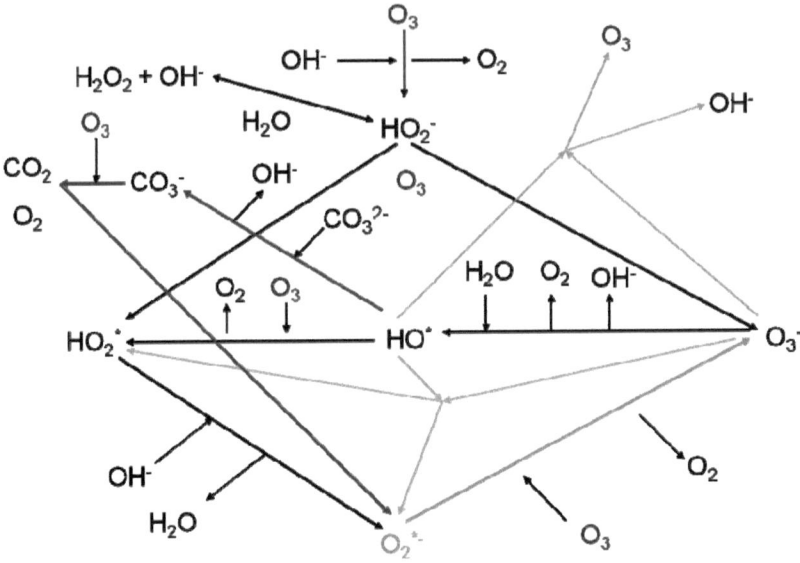

Figure 18 - Mechanism of O_3 decomposition at pH > 10

These proposed decomposition mechanisms (Figure 16 - Figure 18) have one thing in common, ozone generates radicals. Of all these radicals the hydroxyl radical ·OH is the most widely known and investigated as it plays an important role in cellular and atmospheric chemistry. Its standard potential of 2.8 V is even higher than that of molecular O_3 which is 2.07 V, making OH-radicals a very reactive but unselective species. The superoxide radical ·O_2 is a surprising species as it reacts in SET (single electron transfer) reactions as a reducing agent. It is less reactive but is highly selectivity towards O_3. The third radical worthy of mention is the hydroperoxide radical HO_2· as it also plays a role in cellular reactions.

These radicals from the O_3 decomposition in water open up completely new reaction patterns for DI/O_3 leaving the classical ozonolysis behind and coming up with radical reactions instead. The key features of such reaction patterns for radicals are radical addition (Figure 19), radical substitution (Figure 20) and hydrogen abstraction (Figure 21) [18].

Theory – Ozone

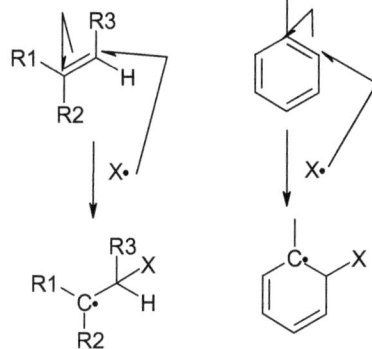

Figure 19 - Mechanism of radical addition

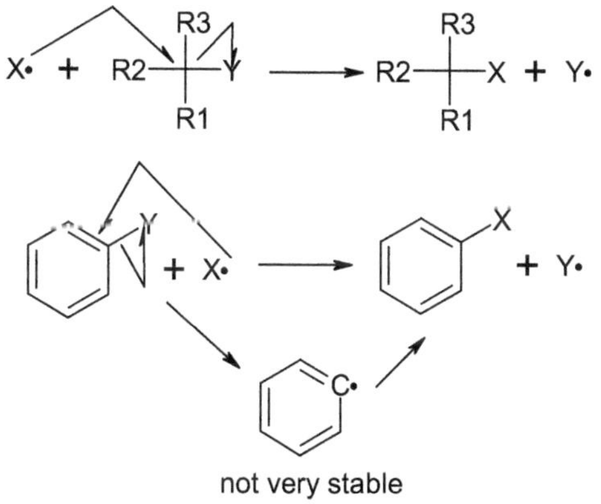

Figure 20 - Mechanism of radical substitution

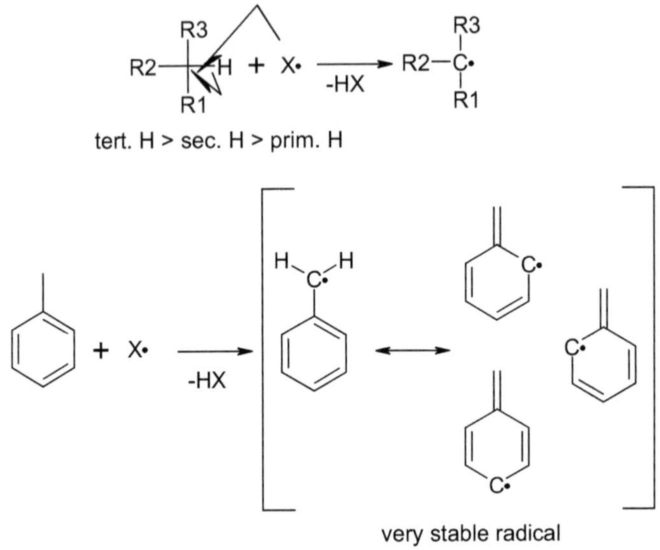
R3
R2
H
+ X•
-HX
R2–C•
R1
tert. H > sec. H > prim. H
+ X•
-HX
H
C•
H
C•
C•
C•
very stable radical

Figure 21 - Mechanism of hydrogen abstraction

Such radical reactions also open up new ways of dealing with modern DUV-resists which are often of the PHS-type (described under 2.1) and therefore have no unsaturated bonds in the polymer backbone. For these saturated structures radicals may offer the best solution for the breakdown of the backbone.

2.3. Radical determination

As explained in the foregoing paragraph the interaction of O_3 with the water it is dissolved in produces radicals which may be useful for the stripping of DUV resists. A method for their detection and quantification would enable the study of the possible influence of additives on their amount. The best and only direct way of radical detection and differentiation is electron spin resonance (ESR). However, as the lifetime of the radicals was too short and the ESR spectrometer in our laboratories was not in the immediate vicinity of the ozone generator, all samples arrived too late at the spectrometer for the detection of signals. This combined with the problems of quantification with ESR only possible by complex sample pretreatment involving the trapping of the highly reactive, unstable OH-radicals and their conversion to stable radicals makes ESR not suited for this study.

2.3.1. The DDL-method

As no method was available for the direct detection and quantification of the radicals an indirect approach was chosen whereby the radicals were trapped by a chemical reaction and converted into stable compounds that could be detected and quantified. Formaldehyde was chosen as the stable molecule. For detection it can be converted to the yellow dye DDL (3,5-diacetyl-1,4-dihydrolutidine) by the Hantzsch [19] reaction and the DDL quantified via its UV/Vis absorbance at 405 nm. This conversion is achieved by the reaction of the formaldehyde as the trapping product with acetylacetone and ammonia (Figure 22).

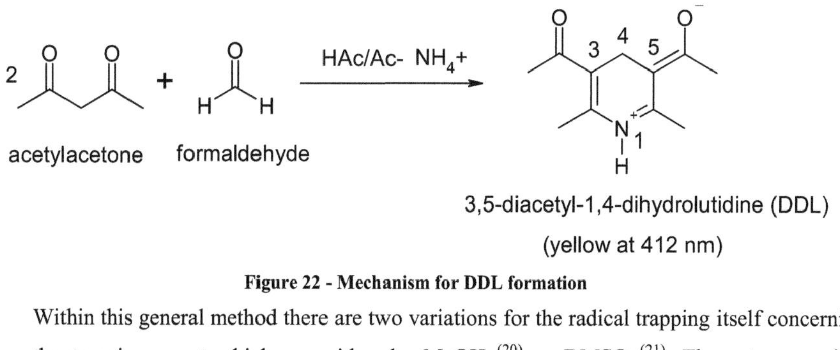

3,5-diacetyl-1,4-dihydrolutidine (DDL)
(yellow at 412 nm)

Figure 22 - Mechanism for DDL formation

Within this general method there are two variations for the radical trapping itself concerning the trapping agent which can either be MeOH [20] or DMSO [21]. These two possible variations offer the scope for a direct cross comparison of the results.

As for every indirect quantitative method a calibration is required it is achieved in this case by using H_2O_2 decomposed by UV-light radiation of 310 nm as the source of defined amounts of radicals. The calibration can either be done as a direct calibration with a reference curve for decomposed H_2O_2 or, better still, by standard addition of H_2O_2. The standard addition is assumed to be the better calibration as it takes into consideration the actual conditions in the sample such as the pH-value which might have an influence on the trapping efficiency of both the trapping agents.

$$H_2O_2 \xrightarrow{h*\nu} 2^{\cdot}OH$$

Figure 23 - H_2O_2 decomposition by UV light

This method involves 2 parallel variations the results of which can be cross checked while each requires an external reference in the form of a calibration curve. A third straightforward method that would serve to confirm the results obtained by radical trapping was deemed useful, if not necessary. Iodometric titration is a self-contained method but not selective for radicals however in combination with UV/Vis. spectroscopy for O_3-selectivity it appeared to fulfil this need.

2.3.1.1. Trapping with MeOH

When MeOH is the trapping agent the ·OH radical is converted in a 1:1 ratio to CH_2O.

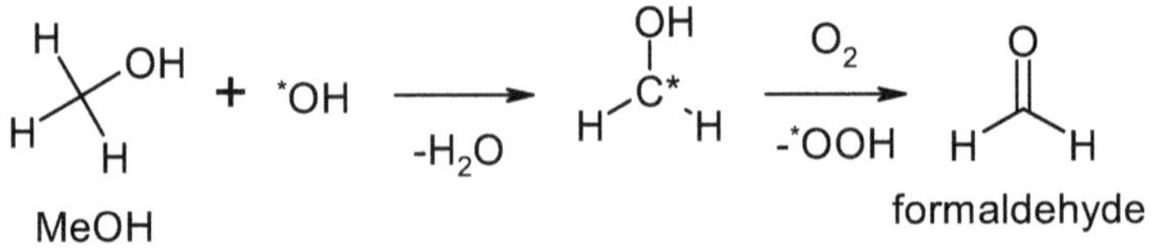

Figure 24 - Mechanism of radical trapping with MeOH

Although this reaction looks simple problems might arise with the practical application as there are known side reactions producing additional radicals. These side reactions occur in acidic and neutral (Figure 25) as well as in alkaline media (Figure 26). They are known as advanced oxidation processes (AOPs) [22]. Their occurrence should influence the slope of the calibration curve so that it is steeper than that obtained theoretically for a simple conversion of ·OH to CH_2O which itself would be twice as steep as the curve obtained with DMSO as the trapping agent.

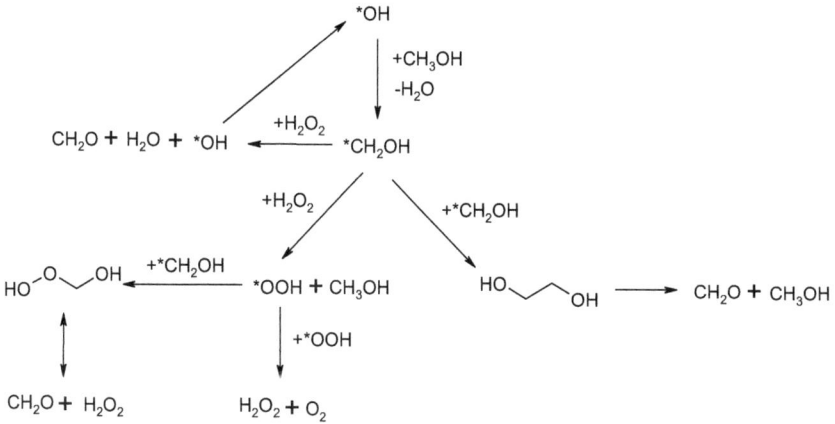

Figure 25 - AOP mechanism in acidic and neutral media

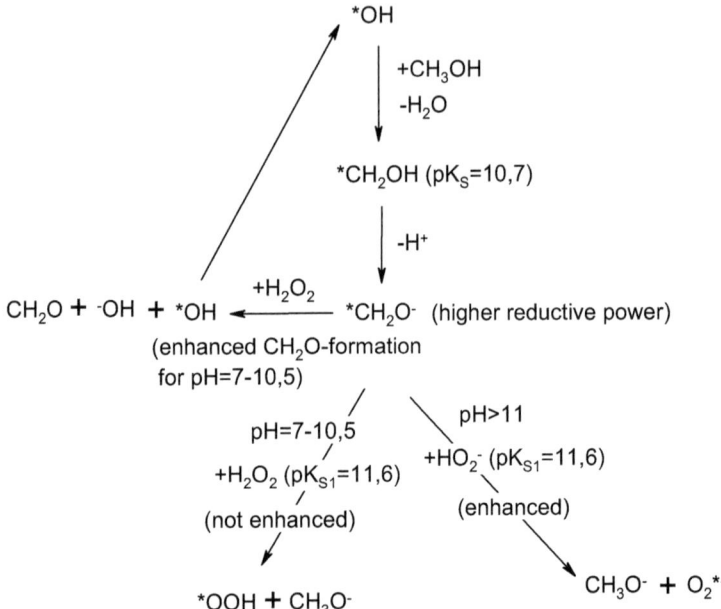

Figure 26 - AOP mechanism in alkaline media

2.3.1.2. Trapping with DMSO

As with MeOH so with DMSO: the conversion ratio of ·OH to CH_2O must be known and can be determined from the trapping mechanism in Figure 27 as 2:1.

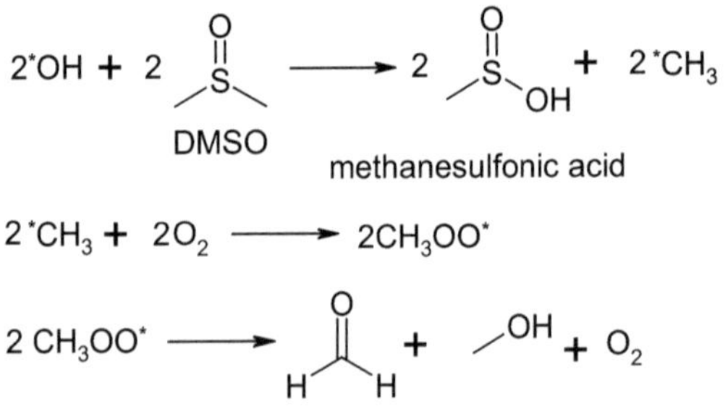

$$2\,^{\bullet}OH + 2\,\,DMSO \longrightarrow 2\,\,\text{methanesulfonic acid} + 2\,^{\bullet}CH_3$$

$$2\,^{\bullet}CH_3 + 2O_2 \longrightarrow 2CH_3OO^{\bullet}$$

$$2\,CH_3OO^{\bullet} \longrightarrow HCHO + CH_3OH + O_2$$

Figure 27 - Mechanism of radical trapping with DMSO

In contrast to the MeOH trapping there are no known undesired side reactions influencing the detection and quantification. Therefore the DMSO trapping should be the more reliable variation of the DDL method. When the influence of the AOPs are ignored and the different conversion ratios of ·OH to CH_2O taken into account, the calibration curve for MeOH as the trapping agent should be twice as steep as that for DMSO.

2.3.2. The combined approach of iodometry and UV/Vis spectroscopy

As mentioned earlier an independent and absolute method for the quantification of the radicals would help verify the results obtained with the DDL method. A combined approach of iodometry and UV/Vis. spectroscopy was adopted. Iodometry was chosen because it is an absolute method without the need of calibration and well suited to determine any kind of oxidizing species including OH-radicals. However, with iodometry the sum of all oxidizing species is determined, and no distinction is made between the OH-radicals to be determined and the O_3 also present in the solution. This called for a selective method for the determination of molecular O_3 which would allow the calculation of the [·OH] as the difference of these two methods. So what would be more obvious than using direct UV/Vis-spectroscopy at 260 nm as applied to the study of the O_3 decomposition and explained earlier (see 2.2).

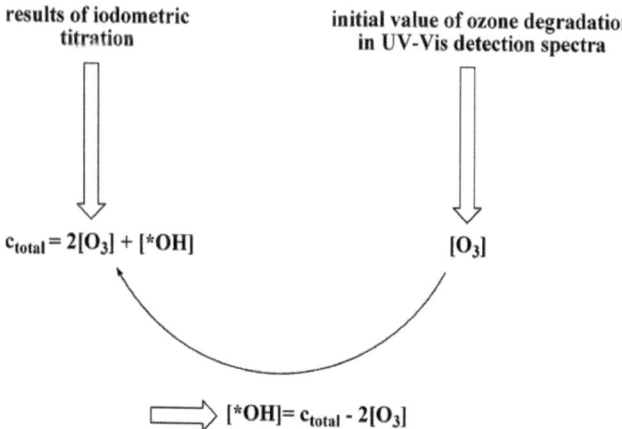

Figure 28 - Schematic for combined iodometric titration and UV/Vis approach

To be able to calculate the [·OH] it is necessary to know the relative thiosulfate ($S_2O_3^{2-}$) consumption of O_3 as well as of ·OH in the iodometric titration. For this purpose the redox equations have to be developed as done in Table III.

Theory – Radical determination

Table III - Redox equations for O_3 and $\cdot OH$ reaction with $S_2O_3^{2-}$ [11]

$\overset{0}{O_3}(O=\overset{+I}{O^+}-\overset{-I}{O^-}) + 2H^+ + 2e^- \longrightarrow \overset{0}{O_2} + \overset{-II}{H_2O}$ $\|E° = 2.07$ V
$\overset{0}{O_3}(O=\overset{+I}{O^+}-\overset{-I}{O^-}) + \overset{-II}{H_2O} + 2e^- \longrightarrow \overset{0}{O_2} + 2\overset{-II}{OH^-}$ $\|E° = 1.24$ V
$\overset{-I}{\cdot OH} + e^- \longrightarrow \overset{-II}{OH^-}$ $\|E° = 2.8$ V
$2I^- \longrightarrow I_2 + 2e^-$ $\|E° = 0.535$ V
$2S_2O_3^{2-} \longrightarrow S_4O_6^{2-} + 2e^-$ $\|E° = 0.1$ V

From these equations the $S_2O_3^{2-}$ consumption can be calculated for O_3 and $\cdot OH$ as done in Table IV and illustrated in Figure 28.

Table IV - $S_2O_3^{2-}$ consumption in iodoemtric titration

$S_2O_3^{2-}$ consumption for O_3:
$\qquad O_3 + 2H^+ + 2e^- \longrightarrow O_2 + H_2O$
$\qquad 2I^- \longrightarrow I_2 + 2e^-$
$\qquad O_3 + 2I^- + 2H^+ \longrightarrow O_2 + I_2 + H_2O$
$\qquad I_2 + 2S_2O_3^{2-} \longrightarrow 2I^- + S_4O_6^{2-}$
$\Rightarrow \qquad \dfrac{\text{mol } S_2O_3^{2-}}{2} = \text{mol } O_3$
$S_2O_3^{2-}$ consumption for $^*OH$:
$\qquad \cdot OH + e^- \longrightarrow OH^- \quad \|x2$
$\qquad 2I^- \longrightarrow I_2 + 2e^-$
$\qquad 2\cdot OH + 2I^- \longrightarrow I_2 + 2OH^-$
$\qquad I_2 + 2S_2O_3^{2-} \longrightarrow 2I^- + S_4O_6^{2-}$
$\Rightarrow \qquad \text{mol } S_2O_3^{2-} = \text{mol } \cdot OH$

According to the above equations, 1 mol molecular O_3 consumes 2 mol of $S_2O_3^{2-}$ whereas 1 mol $\cdot OH$ only consumes 1 mol of $S_2O_3^{2-}$. Therefore the $[\cdot OH]$ can be calculated as $[\cdot OH] = c_{total} - 2 \times [O_3]$.

2.4. IR spectroscopy

For a systematic scientific approach to problems in resist stripping it is necessary to understand the chemical composition of the resist as they appear on the wafer and especially how they change during different process steps. For this purpose an in situ non-destructive analysis is required. As all the resists are organic resins containing polar functional groups (phenols, amines, carboxyls etc.) FT-IR spectroscopy can be used for this purpose. It's a fast, none destructive and easily available technique well suited for all polar functional groups. The simplest way to conduct IR measurements is in the transmission mode (Figure 30) whereby the IR beam is passed directly through the sample. In this case the absorbance depends on the thickness of the sample. In the cases studied the resist thickness is only 400 – 700 nm depending on the resist and the process steps involved. Furthermore this thin resist is applied to the Si-wafer substrate of 700 – 1000 µm thickness through which the IR beam has also to pass. Fortunately the relatively thick Si substrate does not pose a problem as crystalline Si is highly transparent in the IR-region with phononic absorptions between 1000 - 500 $\tilde{v}$ being outside the region of organic group absorbance and therefore do not interfere (Figure 29 – black spectrum).

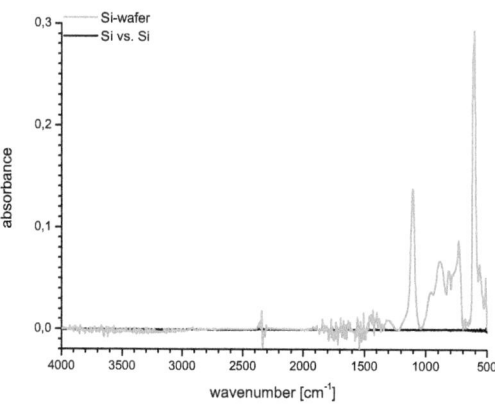

Figure 29 - IR spectra for Si

Furthermore these phononic peaks can be easily removed from the spectra by using the spectra of not just air but air and Si-wafer as the background as shown in Figure 29 - red spectrum. This was the standard method used to subtract the Si background from the spectra in the experiments.

The thinness of the resist might be the bigger problem for the transmission measurements. But this problem can be overcome by increasing the pathway of the light through the sample. The simplest way to do this is to pass the IR beam through the resist at an angle inclined to the surface such that it reflects from the surface of the silicon substrate (see Figure 31). Not only will the pathway through the resist be increased - more than doubled - there will be practically no absorbance by the wafer substrate.

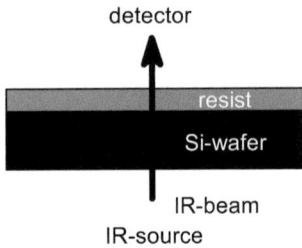

Figure 30 - IR transmission setup

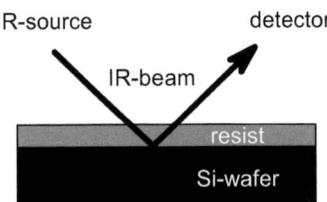

Figure 31 - directed reflection-IR setup

If this method is still not enough to produce peaks large enough to be identified, attenuated total reflection (ATR) may be the solution. In ATR, the IR beam is directed onto an optically dense crystal with a high refractive index at a certain angle allowing multiple total internal reflections. This internal reflectance creates an evanescent wave that extends beyond the surface of the crystal into the sample held in contact with the crystal. The crystal, specially shaped for this purpose, may be germanium, ZnSe or even diamond. The wafer is placed with the resist in contact with the crystal.

The IR beam interacts only with the resist and not the wafer substrate at every point where it is reflected before it exits and is passed on to the detector resulting in an increased absorption signal [23].

Theory – IR spectroscopy

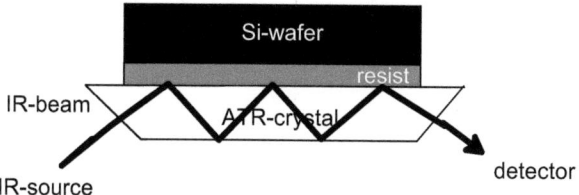

Figure 32 - ATR-IR setup

As initial experiments have shown that the transmission-mode setup is adequate even for the thinnest of the resists studied (~ 400 nm) it was the only mode employed.

Theory – IR spectroscopy

Table V - IR peaks for expected functional groups in the resists[24], [25]

Functional group	IR peaks
Phenol (Ph–OH)	3610 $\tilde{\nu}$; 1200 $\tilde{\nu}$
Ph–H	3100 – 3000 $\tilde{\nu}$; 2000 - 1600 $\tilde{\nu}$; 1600 $\tilde{\nu}$; 1500 $\tilde{\nu}$; 900 - 700 $\tilde{\nu}$
Aromatic ring	1600 - 1500 $\tilde{\nu}$
tert-butyl (C(CH$_3$)$_3$–R)	1380 $\tilde{\nu}$; 1370 $\tilde{\nu}$; 1255 $\tilde{\nu}$; 1210 $\tilde{\nu}$
R$_2$C=CH$_2$	1655 $\tilde{\nu}$
R$_2$CH$_2$ (=CH$_2$)	3080 $\tilde{\nu}$
R–C(=O)–OH	3200 - 2500 $\tilde{\nu}$
R–COO$^-$	1610 - 1550 $\tilde{\nu}$; 1420 - 1300 $\tilde{\nu}$

With the exception of Rohm&Haas UV26 the composition of the resists used in this study is largely unknown. A major problem in using IR spectroscopy to elucidate their structures and understand the changes that take place during the photolithographic process is the fact that the frequencies of group vibrations are not unique but vary with the chemical environment. It is difficult to precisely allocate overlapping peaks to definite functional groups.

To emphasise changes in the resist structures a preferred method is the calculation of differential spectra for the process step being investigated as the appearance or omission of peaks can be used as indicators for a possible mechanism.

2.5. Raman microscopy

Raman spectroscopy is a complementary method to IR as it only interacts with nonpolar but polarisable functional groups. In this study Raman microscopy has only been applied to the implanted DUV248 resist as the ion implantation causes changes not detectable by IR spectroscopy. Changes caused by ion implantation are inhomogeneous in nature – a crust forms on top of the resist film while the bulk of the resist remains unchanged.

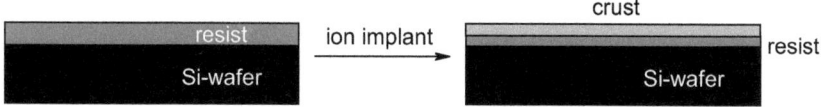

Figure 33 - Scheme for crusted resist

This new 3-layer system is inaccessible to IR spectroscopy in the transmission setup as the thickness of the crusted top resist is only in the region of 20 nm (information from IMEC) compared to the unmodified resist of ~380 nm. Moreover, the crusted top resist more or less consists of highly crosslinked carbon that does not contain the polar groups necessary for IR spectroscopy. This is where Raman microscopy comes in. With Raman microscopy it is possible to detect carbon modifications and distinguish between conventional graphite (1335 cm^{-1}, 1575 cm^{-1}), pyrolitic graphite (1575 cm^{-1}), glassy (1340 cm^{-1}, 1590 cm^{-1}) and diamond like carbon (1332 cm^{-1}) (Figure 35, Figure 36) [26].

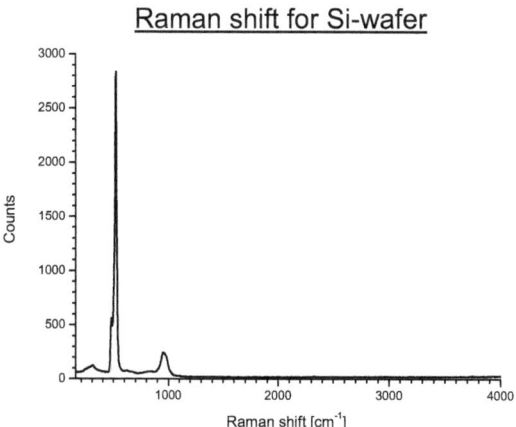

Figure 34 - Raman spectrum of a Si-wafer

Theory – Raman microscopy

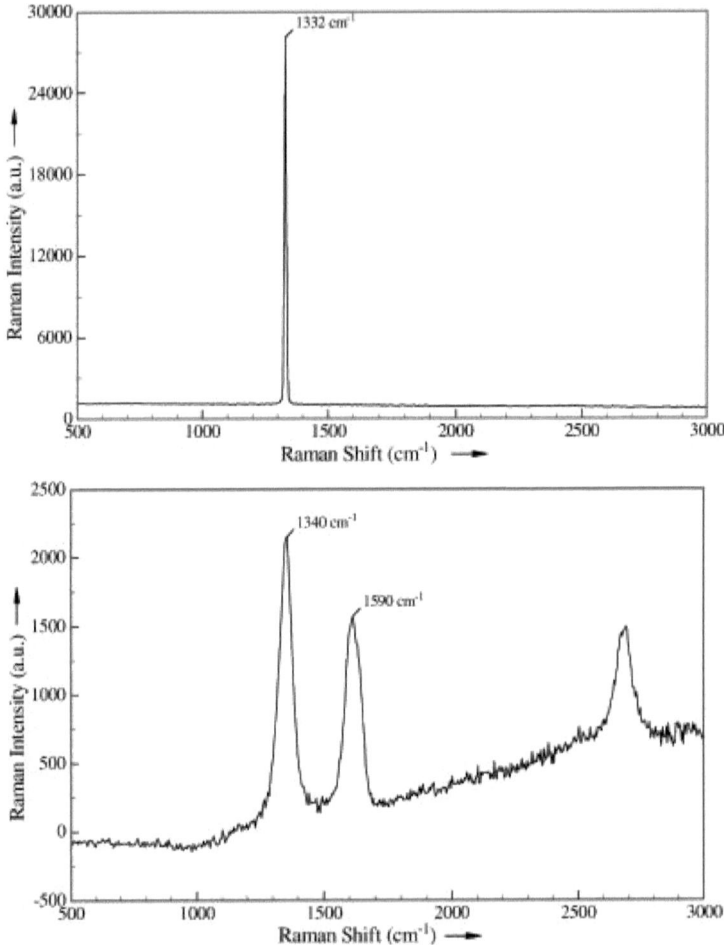

Figure 35 - Raman spectra: gem quality diamond(top), glassy carbon(bottom)

Theory – Raman microscopy

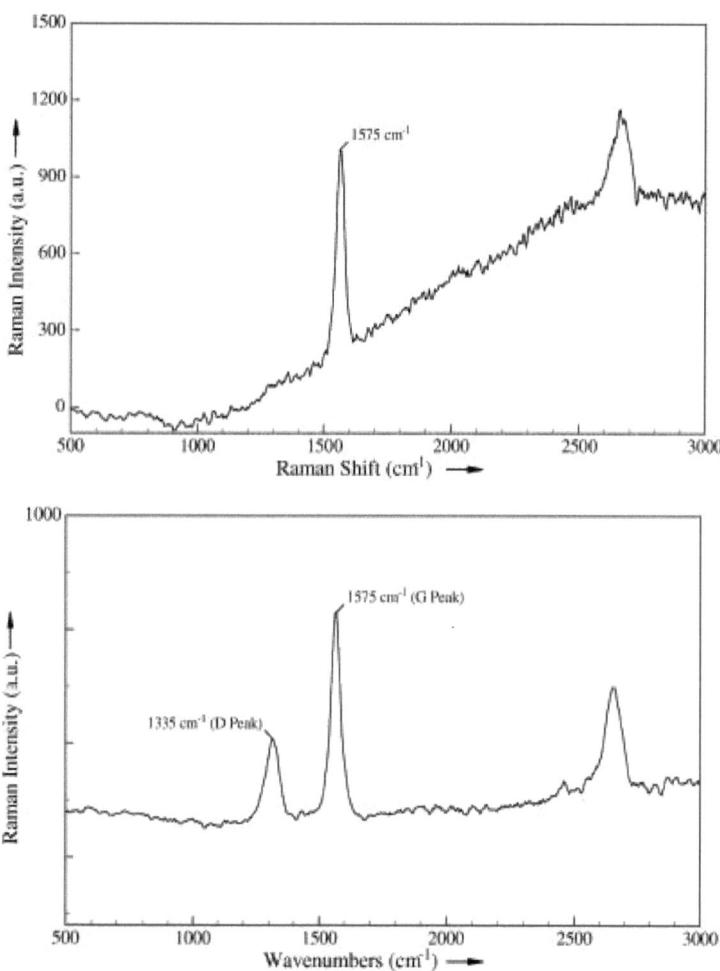

Figure 36 - Raman spectra: highly ordered pyrolitic graphite(top), conventional graphite(bottom)

The use of not only Raman spectroscopy but Raman microscopy as well provides the possibility of only probing the crusted top resist without the interference of the Si-wafer substrate or the unmodified resist layer by the adjustment of the focus of the IR beam stepwise from above the crusted resist layer down to the unmodified resist using low beam power to reduce the excitation depth. These properties make Raman microscopy a tool worthy of investigation for the characterisation of crusted resists.

3. Experimetal Details

Ozone generation
All the ozone used in the experiments was produced from compressed O_2-gas in an Astex SEMOZON 09.2 silent discharge ozone generator, keeping controllable parameters constant. The generator was always operated at 100 % power with an O_2 feed gas flow of 1.5 $\left[\dfrac{L}{min}\right]$ at a working pressure of ~1.8 bar. The O_3-gas thus produced was dissolved in DI-water at RT (~25 °C) - temperature could not be controlled - by a membrane dissolver and then pumped to the sampling location where it was taken as chilled ozone for the experiments. Prior to sampling an initiation time of 30 min was allowed for the generator to reach a steady state ozone concentration of ~1 mM $\hat{=}$ 48 ppm O_3 in solution. This could not be increased by increasing O_2 flow rate or pressure up to the allowed maximum of 2.5 bar. When in operation the DI/O_3 was recirculated through the ozone dissolver at a constant flow rate of 60 $\left[\dfrac{L}{min}\right]$ via a 20 L storage tank to refresh the ozone and maintain this ozone concentration. Excess undissolved ozone was catalytically decomposed before reaching the exhaust. All the necessary components are combined in a professional module constructed for the semiconductor industry by SEZ (Figure 41). To ensure maximum safety the module was completely closed during operation and vented. Both module and exhaust are equipped with sensors which, in the event of an uncontrollable emission of ozone, trigger the shutdown of the whole setup including the generator.

Sampling
Samples were withdrawn under an extractor hood from the recirculation pipe with a needle valve (Figure 42). The samples were taken either with a 300 mL volumetric flask for all experiments needing 150 mL DI/O_3 and transfered into a polypropylene beaker or with a beaker and the necessary amount pipetted with an Eppendorf pipette.

Temperature control
Where temperatures had to be maintained constant a water bath was used (Figure 43). The actual temperature of the solution inside the beaker at water bath temperatures of 50 °C and 90 °C are given in the graph in Figure 45.

Experimental Details – General setup

Resists and wafer preparation

The resists studied in this work go under the trade names JSR KrF M91Y, JSR DUV248 (whereby, according to IMEC, DUV248 is M91Y but with a different viscosity for application on 300 mm wafers) and Rohm&Haas UV26, all with polyhydroxystyrene (PHS) as their basic polymer structure (Figure 37). The chemical nature of the PAC (photo active component) in general or the PAG (photo acid generator) is not known for any of these resists. The nature of the hydroxyl protection groups is only known for Rohm&Haas UV26, being t-BOC (Figure 38). The degree of protection of the hydroxyl groups is not known but is normally about 20 - 30 % depending on the demands for solubility and thermal stability. Further information regarding the composition of the resists was neither available from the manufacturers, nor could it be derived by analytical methods like GC-MS or HPLC. The solvent is either ethyl lactate (Figure 39) or propyleneglycolmethyletheracetate (PMA) (Figure 40) but is of no interest for the studies as it is removed during the soft bake step and therefore not present in the prepared wafers.

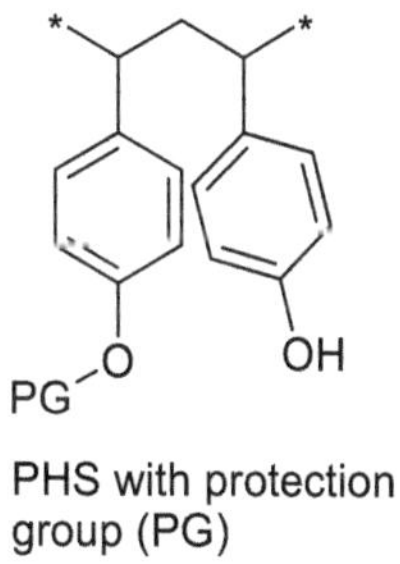

PHS with protection group (PG)

Figure 37 - PHS-PG

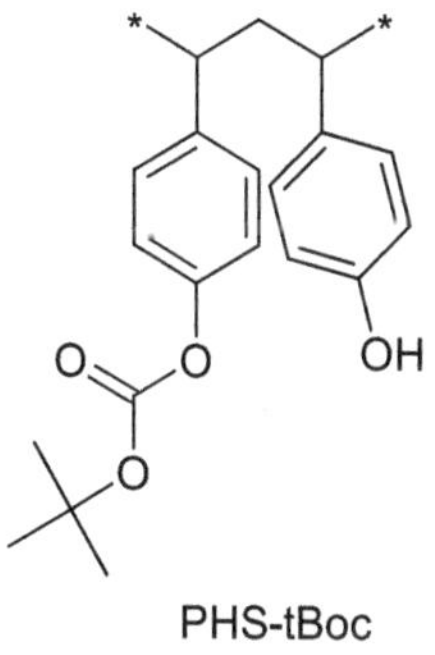

Figure 38 - PHS-tBoc

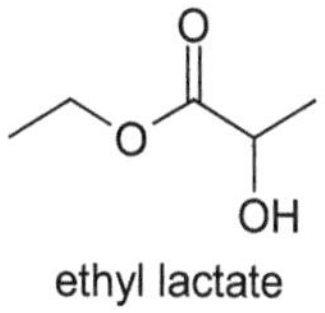

ethyl lactate

Figure 39 - Ethyl lactate

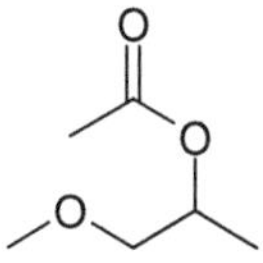

Figure 40 - PMA

Experimental Details – General setup

The wafers were prepared by IMEC using tools and methods as applied in the semiconductor industry. The various steps in the preparation of the resists are given in the table below.

Table VI - Wafer preparation steps

	JSR M91Y	UV26	DUV248
spin coating + soft bake	step 1	step 1	step 1
lithography	step 2	step 2	step 2
post exposure bake/PEB	step 3	step 3	step 3
partial plasma etch	step 4	step 4	
ion implant			step 5

Experimental Details – General setup

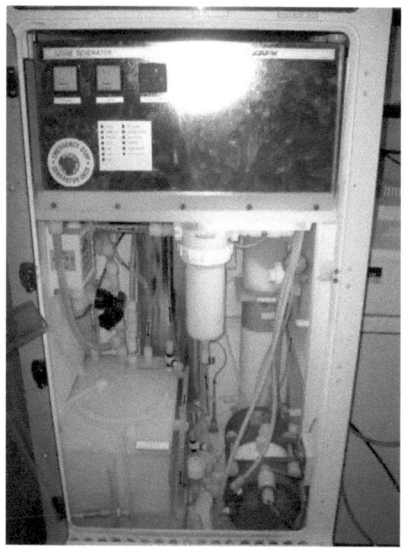

Figure 41 - Ozone module from SEZ

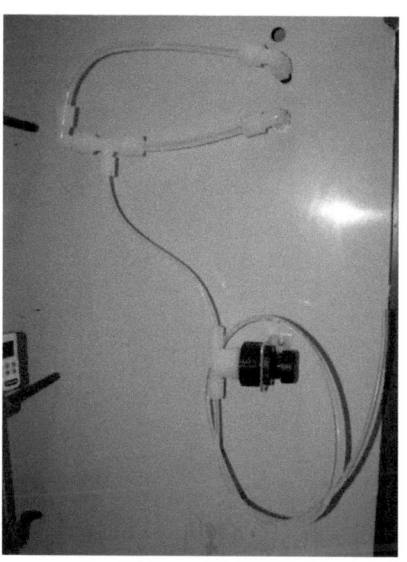

Figure 42 - Sampling setup

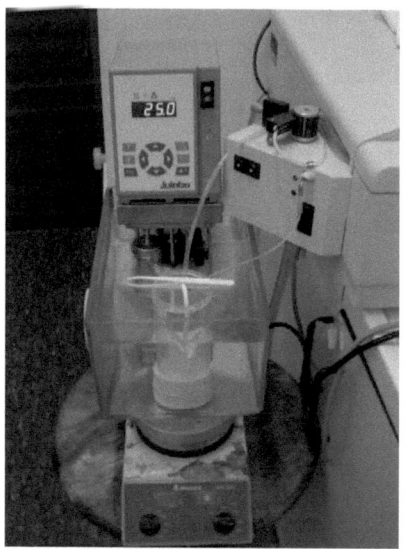

Figure 43 - General setup of the water bath

Figure 44 - UV/Vis-measurement setup

Experimental Details – General setup

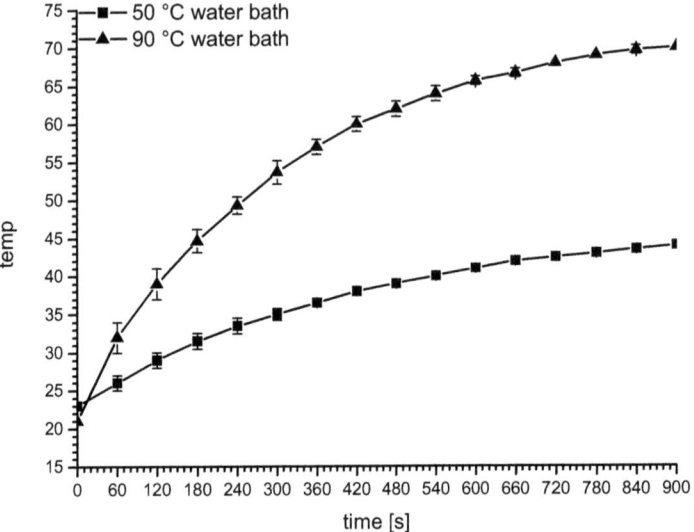

Figure 45 - Temperature of solution in beaker at given water bath temperatures

Experimental Details – Ozone decomposition

3.1. UV/Vis spectroscopic determination of ozone decomposition

All experiments for the determination of the factors affecting ozone decomposition were conducted in 15 mL polypropylene screw capped centrifuge tubes (supplied by Roth) placed in a water bath (Julabo MB-5A heating circulator with open bath, but not equipped with cooling device - Figure 43) for temperature control.

The ozone concentration was monitored directly by an UV/Vis spectrometer at 260 nm using DI-water as reference. Measurements undertaken with DI/O_3 were performed within a liquid flow cell (PerkinElmer Spectroscopy Flow-Through Cell, Semi-Micro/Ultra-Micro with in- and outlet tubes, quartz SUPRASIL®, light path: 10 mm) allowing continuous recirculation as well as stopped flow (Figure 44). For measurements performed under recirculation the flow rate of DI/O_3 was 14 $\left[\dfrac{mL}{min}\right]$ with a liquid volume of 6 mL in the complete circuit and 160 µL in the flow cell. For stopped flow measurements the cell was purged with the solutions under the same parameters for 45 s before the flow was stopped.

The samples always consisted of DI/O_3 plus a pH-altering additive to give a constant sample volume of 10 mL. The amount of additive used in each case was the lowest reasonable volume necessary for the solution to attain the required pH. The additives were chosen to vary the pH-value from acidic to alkaline. The pH-value was determined with a WTW pH90 pH-electrode. To study the influence of the chemical nature of the additives at the same pH different substances were chosen and compared at selected pH-values (Table VII). CO_3^{2-} and HCO_3^{2-}, in particular, were chosen to test their effectiveness as radical scavengers as reported by literature [27].

Experimental Details – Ozone decomposition

Table VII - Additives for ozone decomposition studies

pH	additive	additive [µL]	additive concentration in 10 mL sample $\left[\dfrac{mol}{L}\right]$
1	96 % H_2SO_4 18 $\left[\dfrac{mol}{L}\right]$	120	0.216
	85 % H_3PO_4 14.6 $\left[\dfrac{mol}{L}\right]$	500	0.73
2.3	50 % HF 28.16 $\left[\dfrac{mol}{L}\right]$	500	1.41
3.5	/	/	/
3.6	1 M $KHCO_3$ 9.6 % H_2SO_4 1.8 $\left[\dfrac{mol}{L}\right]$	50 12	0.005 H_2CO_3
7	85 % H_3PO_4 14.6 $\left[\dfrac{mol}{L}\right]$ /H_2O 1:1 with NaOH neutralised to pH=7	500	0.243 $H_2PO_4^-$/HPO_4^{2-}
7.6	1 M $KHCO_3$	50	0.005 CO_3^{2-}
9	28 % NH_4OH 1/150 0.1	500	0.005
	0.1 M NaOH	55	0.00055
	0.1 M KOH	55	0.00055

Experimental Details – Ozone decomposition

The influence of all the additives listed in Table VII were studied with recirculation and stopped flow in quartz flow cell at RT and 50 °C in the tempered water bath as well as in standard quartz cells without purging. From the decomposition curves plotted half life times $\tau_{1/2}$ were directly determined. Using the equation

$$n = 1 + \frac{\lg(\tau_1) - \lg(\tau_2)}{\lg[(A_2)_0] - \lg[(A_1)_0]} \quad (28)$$

where τ_1 and τ_2 are the half life times, $(A_1)_0$ and $(A_2)_0$ are the corresponding starting concentrations, a formal reaction order n is calculated. When it was closer to being a first order reaction ln [O_3] was plotted against t; when it was more of a second order reaction $\frac{1}{[O_3]}$ was plotted against t and the rate constant k calculated as the slope of the linear regression of the plot according to the rate laws.

first order reaction: $\dfrac{d[O_3]}{dt} = -k \times [O_3] \rightarrow \ln[O_3] = -k\left[\dfrac{1}{s}\right] \times t[s] + \ln[O_{3(t=0)}]$ (29)

second order reaction: $\dfrac{d[O_3]}{dt} = -k \times [O_3]^2 \rightarrow \dfrac{1}{[O_3]} = k\left[\dfrac{L}{mol \times s}\right] \times t[s] + \dfrac{1}{[O_{3(t=0)}]}\left[\dfrac{L}{mol}\right]$ (29)

Experimental Details – Ozone decomposition

Applying the Arrhenius equation the activation energy for the reaction was calculated for $T_1 = 25\ °C$ (298.15 K) and $T_2 = 50\ °C$ (323.15 K).

$$\text{Arrhenius equation: } k = A \times e^{-\frac{E_A}{RT}} \quad (29)$$

where k = rate constant, A = pre-exponential factor, E_A = activation energy, R = gas constant and T = temperature [K])

$$k_1 = A \times e^{-\frac{E_A}{R \times T_1}} \; ; \; k_2 = A \times e^{-\frac{E_A}{R \times T_2}}$$

$$A = k_1 \times e^{\frac{E_A}{R \times T_1}} \; ; \; A = k_2 \times e^{\frac{E_A}{R \times T_2}} \rightarrow k_1 \times e^{\frac{E_A}{R \times T_1}} = k_2 \times e^{\frac{E_A}{R \times T_2}}$$

$$\ln k_1 + \frac{E_A}{R \times T_1} = \ln k_2 + \frac{E_A}{R \times T_2} \rightarrow \ln k_1 - \ln k_2 = \frac{E_A}{R \times T_2} - \frac{E_A}{R \times T_1} \rightarrow \ln \frac{k_1}{k_2} = \frac{E_A}{R}\left(\frac{1}{T_2} - \frac{1}{T_1}\right)$$

$$\rightarrow \ln \frac{k_1}{k_2} \times R = E_A \left(\frac{T_1}{T_2 \times T_1} - \frac{T_2}{T_1 \times T_2}\right) \rightarrow \ln \frac{k_1}{k_2} \times R = E_A \left(\frac{T_1 - T_2}{T_2 \times T_1}\right)$$

$$\rightarrow \boxed{E_A = \frac{\ln \frac{k_1}{k_2} \times R \times T_1 \times T_2}{T_1 - T_2}} \; ; \; T_2 > T_1$$

The results of all the above calculations are set out in Table XIII.

3.2. Radical determination

3.2.1. The DDL-method

The DDL method of radical determination was evaluated for its suitability with two different trapping reagents MeOH and DMSO. Both trapping alternatives basically follow the same procedure. For the conversion of the CH_2O formed to DDL a detection solution according to Hantzsch (19) was prepared consisting of 25 g NH_4Ac + 3 mL 100 % HAc + 0.2 mL acetylacetone made up to 100 mL with DI-water.

For both trapping variations the time needed for the complete conversion of CH_2O as well as the stability of the DDL formed were first determined. 20 µL, 80 µL, 140 µL of 36 % CH_2O were each made up to 1 mL to give 3 different concentrations to cover the whole working range of concentrations that would be required later (see Table VIII). The solutions were placed in a water bath maintained at 50 °C and the absorbance of the DDL formed after addition of the Hantzsch reagent was measured at regular intervals (Figure 55). For the calibration curve ($CH_2O \rightarrow$ DDL) standard samples were prepared from 36 % CH_2O and diluted progressively to $1/1000^{th}$ of the standard sample concentration to give a series covering again the range of concentrations required later.

Finally the efficiency of the radical trapping mechanism was determined with calibration curves which involved the entire reaction from trapping to the conversion of CH_2O to DDL. The calibration curves were obtained using 3 different methods: direct measurement in dilute solutions, direct measurement in concentrated solutions and by standard addition to the DI/O_3 sample.

Table VIII - Dilution series for CH_2O

procedure	$[CH_2O] \left[\dfrac{mol}{L}\right]$
0, 20, 40, 60, 80, 120, 140 µL 36 % CH_2O made up to 1 mL with DI-water	0, 0.259, 0.518, 0.777, 1.036, 1.295, 1.554, 1.813
100 µL taken from each of the above and made up to 1 mL	0, 0.0259, 0.0518, 0.0777, 0.1036, 0.1295, 0.1554, 0.1813
10 µL taken from each of the above and made up to 1 mL	0, 2.59×10^{-4}, 5.18×10^{-4}, 7.77×10^{-4}, 1.036×10^{-3}, 1.295×10^{-3}, 1.554×10^{-3}, 1.813×10^{-3}

Experimental Details – Radical determination

Table IX - Composition of the Hantzsch reagent

NH$_4$CH$_3$COO	acetylacetone 100 % C$_5$H$_8$O$_2$	CH$_3$COOH 100 %
77.08 $\left[\dfrac{g}{mol}\right]$	100.12 $\left[\dfrac{g}{mol}\right]$	60.05 $\left[\dfrac{g}{mol}\right]$
	density: 0.975 $\left[\dfrac{g}{mL}\right]$ = 975 $\left[\dfrac{g}{L}\right]$	density: 1.05 $\left[\dfrac{g}{mL}\right]$ = 1050 $\left[\dfrac{g}{L}\right]$
	conc.: 9.74 $\left[\dfrac{mol}{L}\right]$	conc.: 17.49 $\left[\dfrac{mol}{L}\right]$
25 g	0.2 mL	
0.324 [mol] = 3.24 $\left[\dfrac{mol}{L}\right]$ in soln.	0.002 [mol] = 0.02 $\left[\dfrac{mol}{L}\right]$ in soln.	0.52 $\left[\dfrac{mol}{L}\right]$ in soln.
6.5x10^{-3} [mol] / 2 mL	4x10^{-5} [mol] / 2 mL	

formaldehyde CH$_2$O 36 %	methanol CH$_3$OH 100 %	DMSO C$_2$H$_6$SO 100 %
30.03 $\left[\dfrac{g}{mol}\right]$	32.04 $\left[\dfrac{g}{mol}\right]$	78.13 $\left[\dfrac{g}{mol}\right]$
density: 1.08 $\left[\dfrac{g}{mL}\right]$ = 1080 $\left[\dfrac{g}{L}\right]$	density: 0.79 $\left[\dfrac{g}{mL}\right]$ = 790 $\left[\dfrac{g}{L}\right]$	density: 1.1 $\left[\dfrac{g}{mL}\right]$ = 1100 $\left[\dfrac{g}{L}\right]$
conc.: 12.95 $\left[\dfrac{mol}{L}\right]$	conc.: 24.66 $\left[\dfrac{mol}{L}\right]$	conc.: 14.08 $\left[\dfrac{mol}{L}\right]$

All radical trapping experiments were conducted in the same type of centrifugal tubes used for the ozone decomposition studies and screw capped to avoid volume losses.

Table X - [·OH] derived from H$_2$O$_2$ decomposition

μl H$_2$O$_2$	[·OH]	μl H$_2$O$_2$	[·OH]	μl H$_2$O$_2$	[·OH]
10	0.2x10^{-3}	40	0.8x10^{-3}	80x10^{-3}	1.6x10^{-3}
20	0.4x10^{-3}	60	1.2x10^{-3}	100x10^{-3}	2x10^{-3}

Experimental Details – Radical determination

3.2.1.1. Trapping with MeOH

<u>Calibration of DDL formed from CH_2O in the presence of an excess of MeOH:</u>
The following procedure was used in order to determine the influence of excess MeOH, on the formation and detection of the DDL formed.

To 1 mL each of the CH_2O samples diluted according to the procedure in Table VIII, 0, 100 μL, 200 μL, 500 μL and 700 μL of 100 % MeOH were added and the samples made up to 8 mL with DI-water. 2 mL of the Hantzsch reagent were added and the solutions placed in a 50 °C water bath for 30 min to ensure complete conversion. The samples were then allowed to cool down to room temperature and the absorbance at 405 nm was recorded against a blank (Figure 56).

To produce calibration curves for the entire reaction of MeOH right up to the detectable DDL, 30 % H_2O_2 as the ·OH standard provider was added to MeOH in predefined amounts and exposed to UV/Vis radiation (using a Hamamatsu Lightningcure™ L8444-02) for the complete decomposition of the peroxide to ·OH. The calibration was performed directly on diluted and concentrated samples and with standard addition.

<u>Calibration of CH_2O-formation with MeOH (diluted):</u>
0, 10 μL, 20 μL, 40 μL, 60 μL, 80 μL and 100 μL of 30 % H_2O_2 were added to 100 μL aliquots of 100 % MeOH (2.466×10^{-4} mol) and made up to 8 mL with DI-water. The samples were subjected to UV/Vis radiation for 5 min, 7 min, 10 min and 15 min to determine the exposure time for complete H_2O_2 decomposition. 2 mL of Hantzsch reagent were added to each of the solutions which were then placed in a 50 °C water bath for 30 min to allow the DDL to develop. The absorbance was recorded at 405 nm against the blank sample without H_2O_2 addition to compensate for effects from the irradiated water (Version 1 in Figure 57).

Experimental Details – Radical determination

Calibration of CH_2O-formation with MeOH (concentrated):
The same aliquots of 30 % H_2O_2 were added to 100 μL of 100 % MeOH (2.466×10^{-4} mol) but made up to 1 mL and subjected to UV/Vis radiation for the same lengths of time as described in the foregoing procedure. The solutions were then made up to 8 mL with DI-water, 2 mL of Hantzsch reagent added, the solution placed for 30 min in a water bath and the absorbance recorded at 405 nm against a blank (Version 2 in Figure 57).

Calibration of CH_2O-formation with MeOH (standard addition):
To a separate series of solutions with the same concentrations of 30 % H_2O_2 and 100 μL of 100 % MeOH (2.466×10^{-4} mol) and made up to 1 mL as described in the foregoing experiment, were added 5 mL of a sample of unknown concentration. From here on the procedure was identical to that described above (Version 3 in Figure 58).

3.2.1.2. Trapping with DMSO

Calibration of DDL formed from CH_2O in the presence of an excess of DMSO:
The procedure was identical to that used for the calibration of DDL formed from CH_2O in the presence of an excess of MeOH using 45 μL of 100 % DMSO in place of 100 μL of 100 % MeOH (Figure 60).

Calibration of CH_2O-formation with DMSO (diluted):
The calibration was identical to the corresponding calibration with MeOH in the diluted version only replacing the MeOH by 45 μL of 100 % DMSO (Version 1 in Figure 61).

Calibration of CH_2O-formation with DMSO (concentrated):
The calibration was identical to the corresponding calibration with MeOH in concentrated version only replacing the MeOH by 45 μL of 100 % DMSO (Version 2 in Figure 61).

Experimental Details – Radical determination

Calibration of CH_2O-formation with DMSO (standard addition):

The calibration was identical to the corresponding calibration with MeOH by standard addition only replacing the MeOH by 45 µL of 100 % DMSO (Version 3 in Figure 62).

3.2.2. The combined approach of iodometry and UV/Vis spectroscopy

For the iodometric part in this combined approach 100 µL DI/O_3 were added to 10 mL of a 0.1 M KI solution acidified (pH < 2) with 2 mL 7 N H_2SO_4 and left for 0 min, 5 min and 10 min to determine the time needed for the quantitative detection of all oxidizing compounds in the sample. The I_2 formed was then back titrated with 0.1 N $S_2O_3^{2-}$ (thiosulfate) to determine the total amount of oxidizing components in the sample. In accordance with the theory the amount of molecular O_3 was directly determined by UV/Vis-spectroscopy at 260 nm and the amount of $S_2O_3^{2-}$ consumed by the O_3 subtracted from the total $S_2O_3^{2-}$ consumption.

The influence of the application of $(NH_4)_6Mo_7O_{24}$ as a catalyst for faster and quantitative I^- oxidation is tested by the application of two drops of saturated catalyst soln. directly after the acidification (Table XVI).

The results obtained from all the methods employed were compared for pure DI/O_3 as the reference in Table XVII.

3.3. Resist characterization with IR spectroscopy

Infrared spectra were recorded directly of the wafer material (wafer fragments), with the resist as applied still on it, as well as on wafer material at relevant stages of processing and stripping. All the spectra were recorded in transmission against a clean wafer to cancel phononic vibrations of the Si lattice and under an argon atmosphere to exclude water vapour and CO_2 both of which have strong absorptions in the infrared region. Differential spectra were derived from the spectra recorded and enabled peak assignment and observation of the changes in the functional groups, highlighting the changes in the resist structure at various stages of processing and stripping. The results were compared with reference data Chemistry WebBook from NIST (National Institute of Standards and Technology) [30], the SDBS from AIST (National Institute of Advanced Industrial Science and Technology) [31] or with spectra calculated with TeleSpec [32] from the university of Erlangen for assumed PHS-structures with the most common hydroxyl protection groups. Peaks were assigned to functional groups be referring to spectroscopy databases [33] and classical textbooks like "Spektroskopische Methoden in der organischen Chemie" by Hesse, Meier, Zeeh [24].

3.4. Resist characterization with Raman microscopy

Raman measurements were conducted only on the implanted JSR DUV248 resist with a Raman microscope (Microraman Renishaw 1000). The excitation is performed by a laser at 633 nm with 10 % of its available power (5 mW) to avoid any undesired effects on the resist. The scan speed was kept constant at 60 s per scan. For the resist overview the complete available spectral range (150 - 4000 $\tilde{v}$) was recorded and compared to the spectrum of a patterned resist free area of the wafer (Figure 74, Figure 75).

To study the top crusted layer of the resist which was ~20 nm for As $10^{15} \left[\frac{1}{cm^2} \right]$ 5 keV – (information from IMEC) and to distinguish between this crust and the unmodified resist a depth profile was made by adjusting stepwise the focus of the laser from the top of the resist to below it (Figure 76, Figure 77). The resulting spectra were compared to references [26] for different carbon variations.

3.5. Resist stripping

For the majority of the stripping experiments a 300 mL polypropylene beaker was filled with 150 mL DI/O$_3$ (1 mM ≙ 48 ppm O$_3$). The additives are added according to the decomposition studies (15 times the amount used for the 10 mL samples - Table VII). The wafer pieces (about 1x1 cm) were in a teflon holder and immersed completely in the solutions. The beaker was then placed in the tempered water bath (RT, 50 °C, 90 °C) and the solution stirred at 400 rpm. As the DI/O$_3$ did not have the same initial temperature as the water bath, the temperature development in the beaker was recorded (Figure 45). During the stripping the ozone concentration was monitored with UV/Vis at 260 nm by continues recirculation through a flow cell (Figure 43; Figure 46). The solutions were renewed when the O$_3$-concentration had dropped to half its initial value.

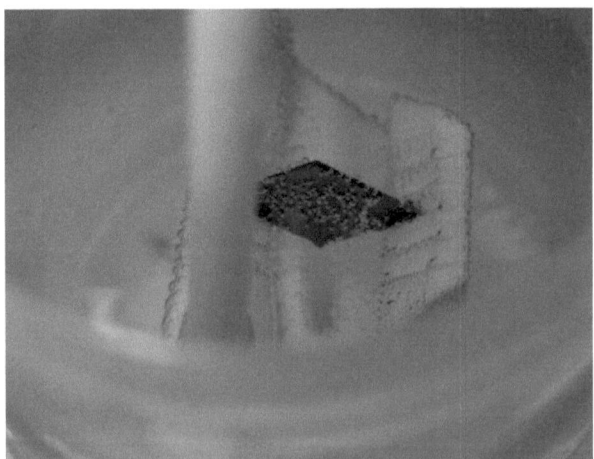

Figure 46 - Sample holder for resist stripping in beaker

Experimental Details – Resist stripping

To investigate the effect of UV irradiation as a way to generate radicals out of the DI/O$_3$ solutions the setup is changed to a dispense setup with a continuous DI/O$_3$ flow (right pipe) over the sample which is directly irradiated by an UV-light source 240 - 400 nm at 3500 $\left[\frac{mW}{cm^2}\right]$ (Hamamatsu Lightningcure^{TM} L8444-02 (LC4) – front fibre optic). With an additional pipe (left pipe) it furthermore is possible to mix the DI/O$_3$ with additives in situ on the wafer allowing higher pH's as the O$_3$ is continuously renewed as well as the combination of these additives with UV. The additives (Table XI) are purged at 8.5 $\left[\frac{mL}{min}\right]$ by a peristaltic sipper from a storage tank (Figure 47).

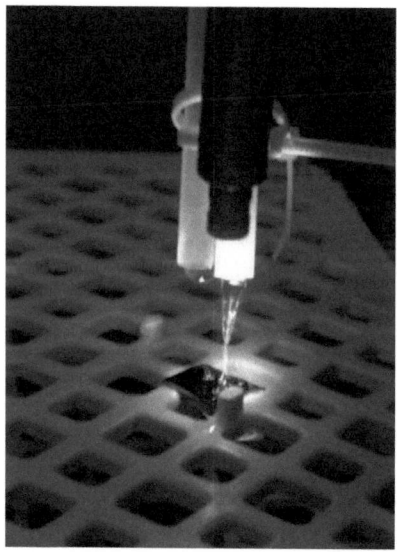

Figure 47 - Setup for dispensed DI/O$_3$ application with additives and UV radiation

Table XI - Additives for in situ mixing

pH of the additive	additive	composition
1	HNO_3	/
1	H_2SO_4	/
6	PO_4^{3-} buffer	/
7	PO_4^{3-} buffer	/
	H_2O_2	30 % H_2O_2 / H_2O 1:1
12	NH_4OH	29 % NH_4OH / H_2O 1:1
12	KOH	/
12.5	KOH	/
13	NH_4OH	29 % pure
13	KOH	/
13	pyrrolidin	> 99 % pure / H_2O 1:1
13.5	KOH	/

In all cases the time needed for complete resist strip is measured (Figure 85, Figure 87, Figure 128 - Figure 131).

When using a hard bake step previous to the stripping it is done on a simple hot plate at 130 °C. For DI/O_3 without any additives as the reference the progress of the resist stripping over time is investigated for the Rohm&Haas UV26 resist by profilometric measurements of the remaining thickness (Figure 89) and inspection of the surface by optical microscopy. The sharp step necessary for profilometric measurements is created by physically removing a part of the resist down to the wafer. This was best achieved by using an adhesive tape for resist pull off.

Also chemical dissolving with organic solvents was tried but only showed poor results in means of a sharp measureable step. Furthermore the pits at the beginning of the resist strip were also sized by the profilometer reflecting the surface roughness (Figure 90, Figure 91). For the resist thinning additional experiments with ellipsometric determination of the resist thickness (Plasmos SD Series; 633 nm laser, 2 layer system, resist: n=1.575, k=0, Si-bulk: n=3.865, k=0.02) [34] were performed but showed no reasonable results aside from the starting thickness.

4. Results and discussion

4.1. UV/Vis spectroscopic determination of ozone decomposition

During the resist decomposition studies the 3 major factors affecting the ozone decomposition have been studied as explained in chapter 3 (3.1) and theorized in chapter 2 (2.2).

Results for half life times $\tau_{1/2}$, reaction orders n and rate constants k are presented in Table XIII. The half life times were derived directly from the recorded plots (Figure 49, Figure 50) and from them the corresponding formal reaction orders and rate constants were calculated. The plots of the 1st order (Figure 51, Figure 52) and 2nd order reactions (Figure 53, Figure 54) graphically illustrate the accuracy with which the decomposition curves conform with the corresponding calculated order.

Factors that affect the decomposition of O_3 in DI water

pH (Figure 49): The fastest decomposition occurred at pH = 9 brought about with NH_4OH, with $\tau_{1/2} \approx 5$ s, the slowest in the pH region of 1 – 4 with $\tau_{1/2} \approx 110$ s, with very little difference between pure DI/O_3 and DI/O_3 with additives, whereby pure DI/O_3 was the most stable of the solutions. Hence, increasing the pH to 9 increased the speed of decomposition by a factor of 22. The pH range studied had no effect on the reaction order n, which is generally ~ 1.

It is noteworthy that the three different pH-dependent decomposition mechanisms for ozone (Figure 16, Figure 17, Figure 18) find their analogy in the decomposition rates for the pH-values 1 - 4, 4 - 7 and above 7 whereby within each pH region there are no significant differences.

Temperature at 50 °C and with continuous recirculation in flow cells (Figure 50): The fastest decomposition also occurred at pH = 9 with NH$_4$OH, $\tau_{1/2} \approx 5$ s, the slowest also in the pH region 1 - 4 but with $\tau_{1/2} \approx 75$ s with little difference between pure DI/O$_3$ and DI/O$_3$ with additives, pure DI/O$_3$ being again the most stable solution. Thus, increasing the temperature from 25 °C to 50 °C can increase the speed of decomposition by a factor of 1.5. The reaction order n, which is generally ~ 1, is not influenced by this increase in temperature. For slower decomposition rates at lower pH levels the temperature effect is stronger than for faster rates at higher pH values. The reason is the time needed for the sample solution to warm up in the tempered water bath (Figure 45). Apart from the effect on the decomposition constant k, higher temperatures also increase outgassing of molecular ozone and of the O$_2$ resulting from its decomposition.

Recirculation, stopped flow and stagnancy at 25 °C produced interesting results. In general the decomposition of the stagnant fluid in the quartz cell was the slowest and with recirculation the fastest. The biggest difference occurred at pH = 1 (DI/O$_3$ + H$_2$SO$_4$) where $\tau_{1/2} \approx 70$ s as opposed to $\tau_{1/2} \approx 5610$ s (Figure 113): an increase by a factor of 80. Even with pure DI/O$_3$, where the effect of additives could be ruled out, the increase in $\tau_{1/2}$ was 35-fold (Figure 119). Values for the stopped flow cell are located somewhere in between. Another interesting effect of no sample recirculation is the fact that in the pH region 1 - 4 the different pH levels could be distinguished by their half life times which was not possible with recirculation. Without recirculation the reaction order n, which is generally ~ 1, increased to 1.6.

CO$_3^{2-}$ or HCO$_3^-$: Formed from atmospheric CO$_2$ dissolved in the solution, they have been reported to act as radical scavengers [27]. This could not be confirmed under the given conditions as in their presence, only possible at pH ≥ 7 according to the Hägg-diagram for H$_2$CO$_3$ (Figure 48) [35], the effect of the alkaline medium outweighs a possible effect as radical scavenger.

$$H_2CO_3 \leftrightarrow H^+ + HCO_3^-; pK_{S1} = 6.5$$

$$HCO_3^- \leftrightarrow CO_3^{2-}; pK_{S2} = 10.4$$

Results and Discussion – Ozone decomposition

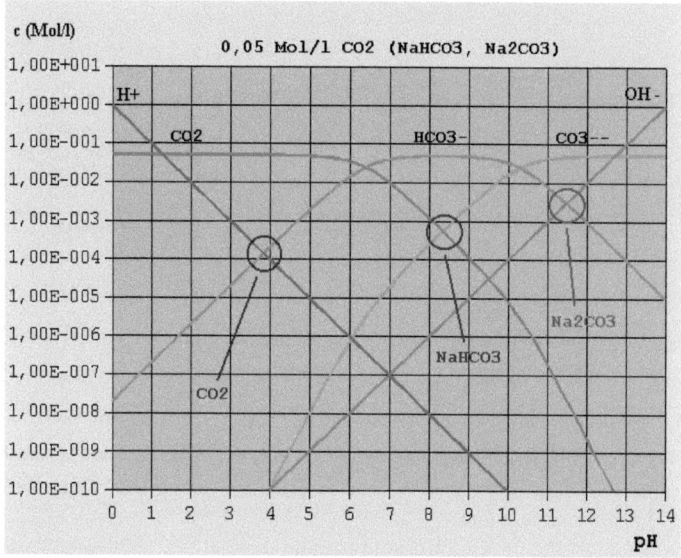

Figure 48 - Hägg-diagram for H_2CO_3

The analysis of the calculated activation energies E_A for pure DI/O_3 indicates an increase from recirculation to no recirculation $E_{A(recirc.)} < E_{A(stopped\ flow)} < E_{A(no\ recirc.)}$ whereas in those cases where additives were used the activation energy for stopped flow is the highest $E_{A(recirc.)} < E_{A(no\ recirc.)} < E_{A(stopped\ flow)}$. In all cases the activation energy for continuous recirculation is the lowest. In DI/O_3 samples with additives the largest increase occurred at pH = 9 with NH_4OH with $\Delta E_A = 153 \left[\dfrac{kJ}{mol}\right]$ as compared to 36 $\left[\dfrac{kJ}{mol}\right]$ for pure DI/O_3 at pH = 3.5.

Results and Discussion – Ozone decomposition

To sum up the results, the physical effect of recirculation and with it the formation of nuclei for outgassing of both O_3 and O_2 from the solution is the strongest influence on the concentration of O_3 in DI-water with $\tau_{1/2}$ increased up to 35-fold in pure DI/O_3.

In DI/O_3 with additives the effect of mixing brought about by recirculation produced an increase in t ½ of up to 80 times. As for the additives the effect produced by altering the pH was the most significant with a massive increase in the decomposition rate in an alkaline medium by a factor of 22. An increase in temperature asserted the least influence – $\tau_{1/2}$ at 50 °C being only 1.5 times more than at 25°C. Due to their alkalinity CO_3^{2-} or HCO_3^- did not produce any change attributable to their property as radical scavengers. Formal reaction orders were close to 1st order in most cases and were not influenced by any of the variable parameters except recirculation. Activation energies over the whole range of variations extended from 13 – 176 $\left[\dfrac{kJ}{mol}\right]$ whereby with NH_4OH at pH 9 alone they ranged from 23 – 176 $\left[\dfrac{kJ}{mol}\right]$.

A detailed comparison of values obtained for the reaction rates k and activation energies E_A with data from other sources has not been undertaken for several reasons. The conditions (water quality, additives, mixing, temperature) under which the values were obtained were either not always the same or they were simply not stated. Published data have been recorded over a wide range of parameters and would make a comparison futile. Table XII therefore is only intended for a quick overview.

Results and Discussion – Ozone decomposition

Table XII - Comparison of experimentally obtained constants for k and E_A with hitherto published values

pH	k (experimental)	k (literature)	E_A (experimental)	E_A (literature)
0.5 – 3			$18-94 \left[\frac{kJ}{mol}\right]$	$59 \left[\frac{kJ}{mol}\right]$ (36)
2 (phosphate-buffered)	at pH = 1 $4.41\times10^{-5} \left[\frac{1}{s}\right]$	$8.3\times10^{-5} \left[\frac{1}{s}\right]$ (37)	/	/
7 (phosphate-buffered)	$1.78\times10^{-4} \left[\frac{1}{s}\right]$	$4.8\times10^{-4} \left[\frac{1}{s}\right]$ (37) $12.8\times10^{-4} \left[\frac{1}{s}\right]$ (38)	/	/
pH~8	$1.26\times10^{-2} \left[\frac{1}{s}\right]$	$2.41\times10^{-2} \left[\frac{1}{s}\right]$ (39)	/	/
pH = 6 - 9	$1.8\times10^{-4} \left[\frac{1}{s}\right]$ - $6.3\times10^{-2} \left[\frac{1}{s}\right]$	$4.72\times10^{-5} \left[\frac{1}{s}\right]$ - $1.55\times10^{-3} \left[\frac{1}{s}\right]$ (40)	/	/
HCO_3^-	$1.972 \left[\frac{L}{mol*s}\right]$	$1.5\times10^7 \left[\frac{L}{mol*s}\right]$ (41)	/	/
CO_3^{2-}	/	$4.2\times10^8 \left[\frac{L}{mol*s}\right]$ (41)	/	/

A more detailed list of reaction constants assigned to different intermediate stages of the ozone decomposition can be found in [42].

Ozone decomposition is sensitive to a host of parameters and it is therefore recommended that the required values be determined for each setup and condition.

Results and Discussion – Ozone decomposition

Table XIII - Results of ozone decomposition studies

	pH = 1					
	H_2SO_4			H_3PO_4		
treatment	recirc flow cell	stopped flow	no recirc quartz cell	recirc flow cell	stopped flow	no recirc quartz cell
$\tau_{1/2}$ [s] 25 °C	70; 80; 80	/	5610	105; 105; 100	/	750; 1590
$\tau_{1/2}$ [s] 50 °C	75; 60; 55	/	1410; 1890; 2610	75; 55; 55	1410; 1440; 1560	1260; 1140; 1050
order n in [O_3] 25 °C	1.1	/	/	1.0	/	2.1
order n in [O_3] 50 °C	0.8	/	1.4	0.8	1.1	0.9
k calculated from n 25 °C	$6.34 \times 10^{-3} \left[\frac{1}{s}\right]$	$1.98 \times 10^{-5} \left[\frac{1}{s}\right]$ $0.0381 \left[\frac{L}{mol*s}\right]$	$1.311 \times 10^{-4} \left[\frac{1}{s}\right]$	$6.67 \times 10^{-3} \left[\frac{1}{s}\right]$	$4.41 \times 10^{-5} \left[\frac{1}{s}\right]$ $0.08782 \left[\frac{L}{mol*s}\right]$	$1.223 \left[\frac{L}{mol*s}\right]$
k calculated from n 50 °C	$1.1 \times 10^{-2} \left[\frac{1}{s}\right]$	$9.41 \times 10^{-5} \left[\frac{1}{s}\right]$ $1624.7 \left[\frac{L}{mol*s}\right]$	$3.367 \times 10^{-4} \left[\frac{1}{s}\right]$	$1.19 \times 10^{-2} \left[\frac{1}{s}\right]$	$4.929 \times 10^{-4} \left[\frac{1}{s}\right]$	$7.74 \times 10^{-4} \left[\frac{1}{s}\right]$
activation energy E_A [kJ/mol] 1st / 2nd	~18 /	50 / 342	30 /	19 /	77 /	/

Results and Discussion – Ozone decomposition

	pH = 2 - 4					
	HF pH = 2.3			pure pH = 3.5		
treatment in flow cell	recirc flow cell	stopped flow	no recirc quartz cell	recirc flow cell	stopped flow	no recirc quartz cell
$\tau_{1/2}$ [s] 25 °C	95; 90; 85	1740; 1950; 2340	3600; 4110	110;95;100	2250; 3450	3450
$\tau_{1/2}$ [s] 50 °C	70; 55; 55	210; 110; 100	1410; 1320; 1620	60; 60; 50	480; 560; 900	420; 520; 780
order n in [O_3] 25 °C	0.9	1.2	1.2	1.0	1.6	/
order n in [O_3] 50 °C	0.8	0.5	1.1	0.9	1.5	1.4
k calculated from n 25 °C	$6.19 \times 10^{-3} \left[\frac{1}{s}\right]$	$3.215 \times 10^{-4} \left[\frac{1}{s}\right]$	$1.777 \times 10^{-4} \left[\frac{1}{s}\right]$	$6.49 \times 10^{-3} \left[\frac{1}{s}\right]$	$1.96 \times 10^{-4} \left[\frac{1}{s}\right]$ $0.795 \left[\frac{L}{mol*s}\right]$	$1.35 \times 10^{-4} \left[\frac{1}{s}\right]$
k calculated from n 50 °C	$1.2 \times 10^{-2} \left[\frac{1}{s}\right]$	$6.12 \times 10^{-3} \left[\frac{1}{s}\right]$	$7.83 \times 10^{-4} \left[\frac{1}{s}\right]$	$1.33 \times 10^{-2} \left[\frac{1}{s}\right]$	$6.993 \times 10^{-4} \left[\frac{1}{s}\right]$ $6.818 \left[\frac{L}{mol*s}\right]$	$8.592 \times 10^{-4} \left[\frac{1}{s}\right]$
activation energy E_A [kJ/mol] 1st / 2nd	21 /	94 /	48 /	23 /	41 / 69	59 /

Results and Discussion – Ozone decomposition

treatment in flow cell	pH = 2 - 4		
	H_2CO_3 pH = 3.6		
	recirc flow cell	stopped flow	no recirc quartz cell
$\tau_{1/2}$ [s] 25 °C	90; 95; 100	/	/
$\tau_{1/2}$ [s] 50 °C	75; 55; 60	420; 570; 1350	450; 600; 870
order n in [O_3] 25 °C	1.1	1.7	/
order n in [O_3] 50 °C	0.8	1.8	1.5
k calculated from n 25 °C	$5.62 \times 10^{-3} \left[\frac{1}{s}\right]$	$2.53 \times 10^{-4} \left[\frac{1}{s}\right]$ $1.33941 \left[\frac{L}{mol*s}\right]$	/
k calculated from n 50 °C	$1.16 \times 10^{-2} \left[\frac{1}{s}\right]$	$5.22137 \left[\frac{L}{mol*s}\right]$	$7.81 \times 10^{-4} \left[\frac{1}{s}\right]$
activation energy E_A [kJ/mol] 1st / 2nd	23 /	/ 44	/

Results and Discussion – Ozone decomposition

pH = 7 - 8

treatment in flow cell	HCO$_3^-$ pH = 7.6			H$_2$PO$_4^-$/HPO$_4^{2-}$ pH = 7		
	recirc flow cell	stopped flow	no recirc quartz cell	recirc flow cell	stopped flow	no recirc quartz cell
$\tau_{1/2}$ [s] 25 °C	65; 55; 55	1440; 2490; 3420	/	50; 50; 50	1110; 2460	2490
$\tau_{1/2}$ [s] 50 °C	50; 45; 55	35; 35; 40	/	40; 35; 25	90; 140; 210	130; 120; 160
order n in [O$_3$] 25 °C	0.9	1.6	/	1	2.1	/
order n in [O$_3$] 50 °C	1.1	1.1	/	0.7	1.6	1.1
k calculated from n 25 °C	$1.26 \times 10^{-2} \left[\frac{1}{s}\right]$	$2.461 \times 10^{-4} \left[\frac{1}{s}\right]$ $1.972 \left[\frac{L}{mol*s}\right]$	/	$1.32 \times 10^{-2} \left[\frac{1}{s}\right]$	$1.555 \left[\frac{L}{mol*s}\right]$	$1.78 \times 10^{-4} \left[\frac{1}{s}\right]$ $0.595 \left[\frac{L}{mol*s}\right]$
k calculated from n 50 °C	$1.89 \times 10^{-2} \left[\frac{1}{s}\right]$	$1.839 \times 10^{-2} \left[\frac{1}{s}\right]$	/	$2.34 \times 10^{-2} \left[\frac{1}{s}\right]$	$2.75 \times 10^{-3} \left[\frac{1}{s}\right]$ $84.3 \left[\frac{L}{mol*s}\right]$	$3.59 \times 10^{-3} \left[\frac{1}{s}\right]$
activation energy E$_A$ [kJ/mol] 1st / 2nd	13 /	138 /	/	18 /	/ 128	96 /

Results and Discussion – Ozone decomposition

		pH = 9				
		NH$_4$OH		KOH		NaOH
treatment in flow cell	recirc flow cell	no recirc quartz cell	recirc flow cell	no recirc quartz cell	recirc flow cell	no recirc quartz cell
$\tau_{1/2}$ [s] 25 °C	3; 4.3; 7.4	15.9; 6; 16.1	38; 90	2490	/	5; 9; 15
$\tau_{1/2}$ [s] 50 °C	4; 6.6; 6	2.6; 3.1; 3	/	15.9; 6; 16.2	/	/
order n in [O$_3$] 25 °C	1.7	1	2.2	/	/	1.8
order n in [O$_3$] 50 °C	1.3	1.1	/	1	/	/
k calculated from n 25 °C	6.27x10^{-2} $\left[\frac{1}{s}\right]$ 4296 $\left[\frac{L}{mol*s}\right]$	8.12x10^{-4} $\left[\frac{1}{s}\right]$	/	/	/	0.824x10^{-2} $\left[\frac{1}{s}\right]$ 4684 $\left[\frac{L}{mol*s}\right]$
k calculated from n 50 °C	0.12668 $\left[\frac{1}{s}\right]$	0.19676 $\left[\frac{1}{s}\right]$	/	/	/	/
activation energy E$_A$ [kJ/mol] 1st / 2nd	23 /	176 /	/	/	/	/

Results and Discussion – Ozone decomposition

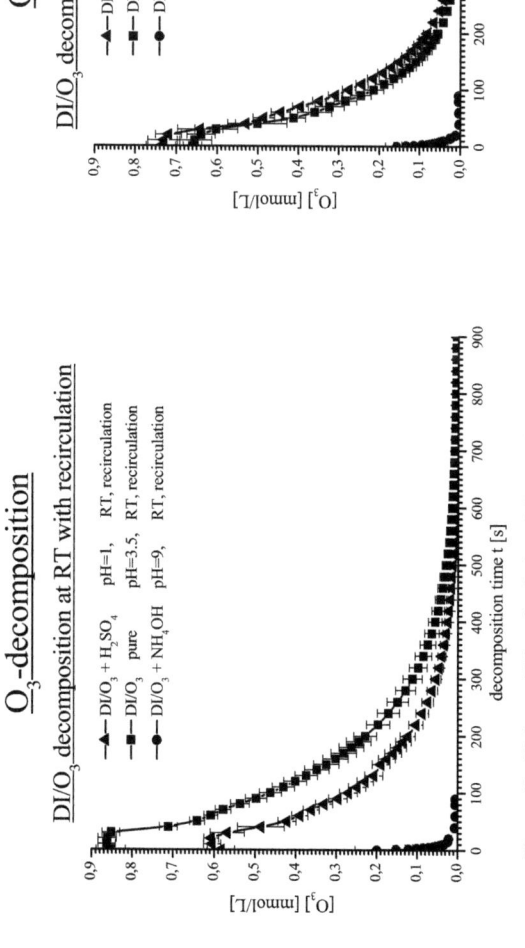

Figure 49 - [O_3] vs. t; RT; recirculation; different pH values

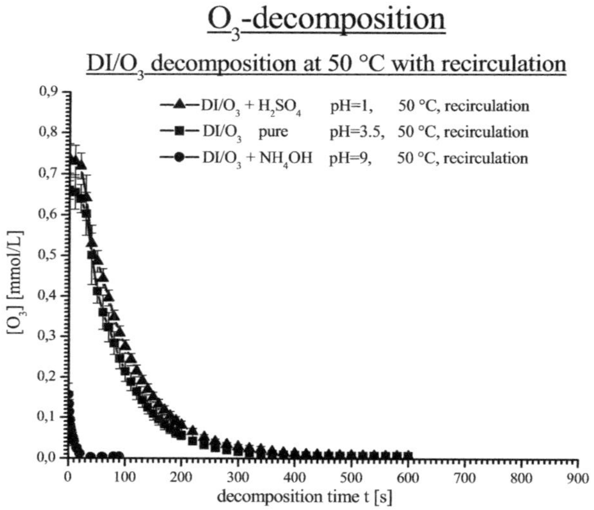

Figure 50 - [O_3] vs. t; 50 °C; recirculation; different pH values

79

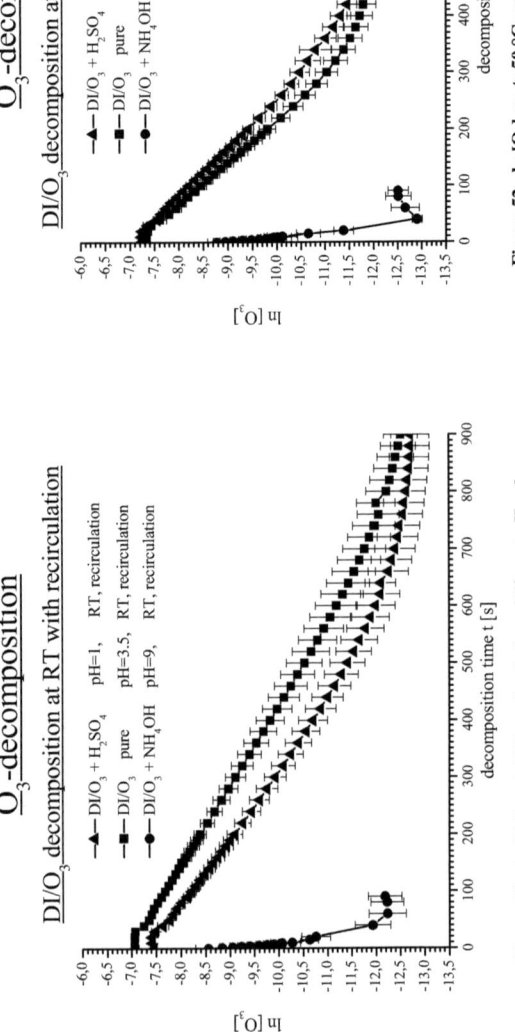

Figure 51 - ln [O₃] vs. t; RT; recirculation; different pH values

Figure 52 - ln [O₃] vs. t; 50 °C; recirculation; different pH values

Results and Discussion – Ozone decomposition

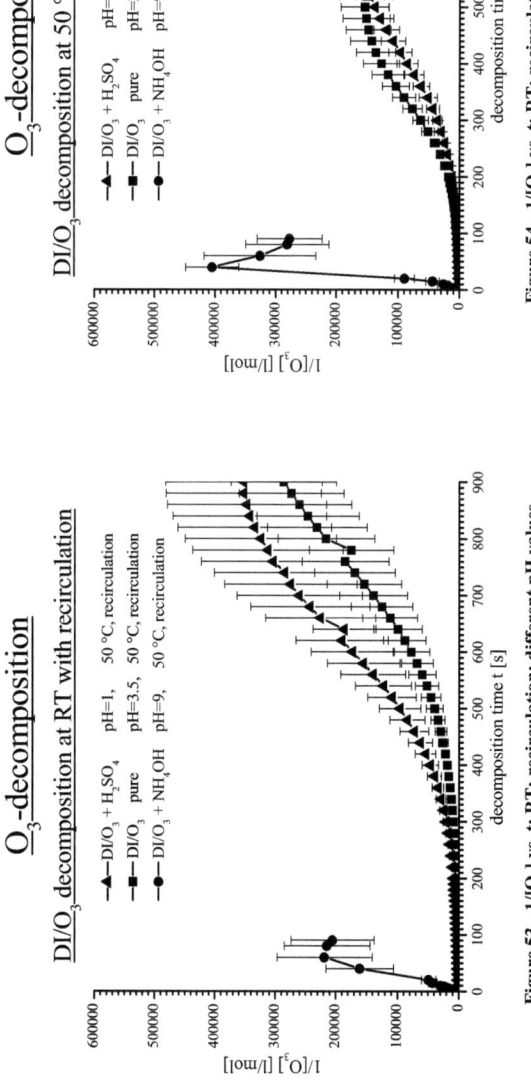

Figure 53 - 1/[O_3] vs. t; RT; recirculation; different pH values

Figure 54 - 1/[O_3] vs. t; RT; recirculation; different pH values

81

4.2. Radical determination

4.2.1. The DDL-method

As explained in the experimental section the first factor to be investigated was the time required for the complete conversion of CH_2O to DDL and the long term stability of DDL under the conditions of the experiment. The absorbance in Figure 55 shows that within the given range 30 minutes at 50 °C was more than enough for complete DDL formation, regardless of CH_2O concentration, and may even be reduced to 10 minutes when necessary. A second positive aspect is that the DDL remained stable for more than 4 hours under these conditions. This wide time range for the DDL formation from 10 - 240 min makes it a very robust, easy to handle method.

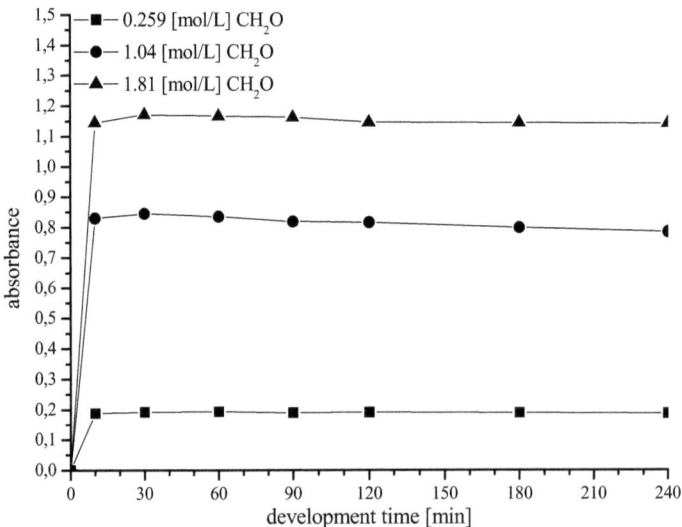

Figure 55 - DDL development and stability

The next step was to develop a practical method to determine the quantitative relationship between the amount of DDL detected and the concentration of the CH_2O it represented. This was achieved, as in most indirect methods of quantification, with a calibration curve.

4.2.1.1. Trapping with MeOH

Where MeOH was the trapping agent it had to be present in excess to trap all the available OH radicals. It was therefore necessary to rule out the possibility that unreacted MeOH interfered by distorting the DDL values obtained. It can be seen from Figure 56 that, regardless of the amount of MeOH used, all plots were linear, the slopes of all the plots were almost identical (777 – 835) and, by comparison with Figure 60, also identical with the calibration plots obtained with DMSO as the trapping agent. All these factors serve to confirm that the unreacted MeOH had no adverse effect on the calibration.

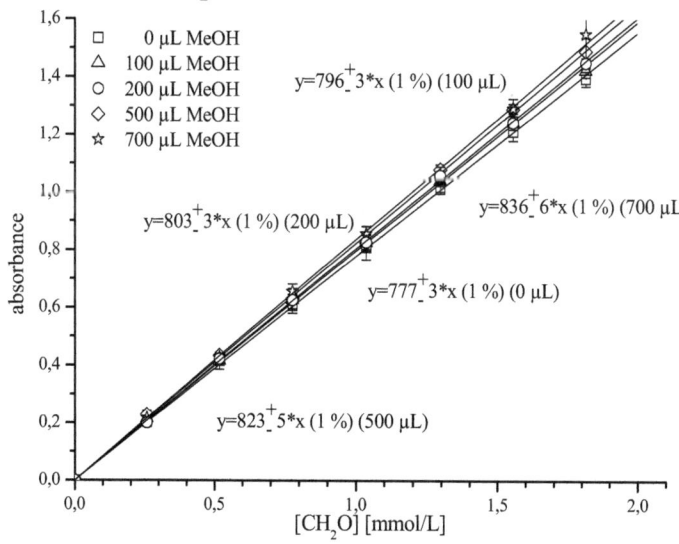

Figure 56 - Calibration of CH_2O conversion to DDL with MeOH in excess

Results and Discussions – Radical determination

Once the reliability of MeOH in this method had been established, a calibration curve to determine the optimum parameters for a complete conversion of OH radicals to CH_2O and further on to DDL as explained in chapter 2 (see 2.3.1.1.) was the next step. Figure 57 shows the plots obtained by direct calibration and Figure 58 those obtained by standard addition. All the plots are linear over the whole range of concentrations studied and for all variations in concentration and method.

Comparing the direct calibration in the diluted and concentrated variations (Figure 57) the slope of the calibration curves for concentrated ones is always steeper than for the diluted ones under the same conditions with the biggest difference at 5 min irradiation time. For the diluted variation the slope increases with the irradiation time (965 → 1315) whereas for the concentrated variation the slope decreases (1455 → 1208). For a 10 min UV-irradiation the difference between the diluted and concentrated variation is minimal (1257 vs. 1315). So a direct calibration in concentrated solutions with 10 min UV-irradiation for H_2O_2 decomposition is the most promising attempt for a direct calibration approach as the curves obtained from the concentrated variation have a lower standard deviation of their slopes.

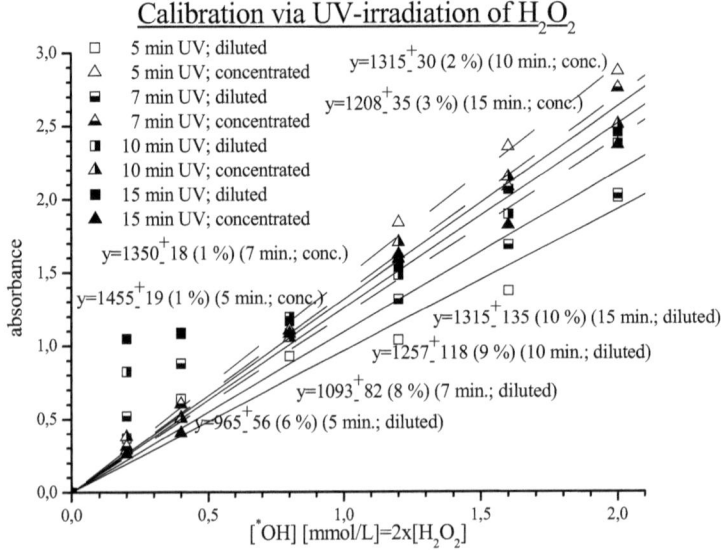

Figure 57 - Calibration of radical trapping with MeOH in diluted & concentrated form

Results and Discussions – Radical determination

Standard addition calibration curves were all linear. The y-offset increased with increasing irradiation times, whereas the slope decreased. The radical concentrations calculated from the plots increased with irradiation time (Table XIV).

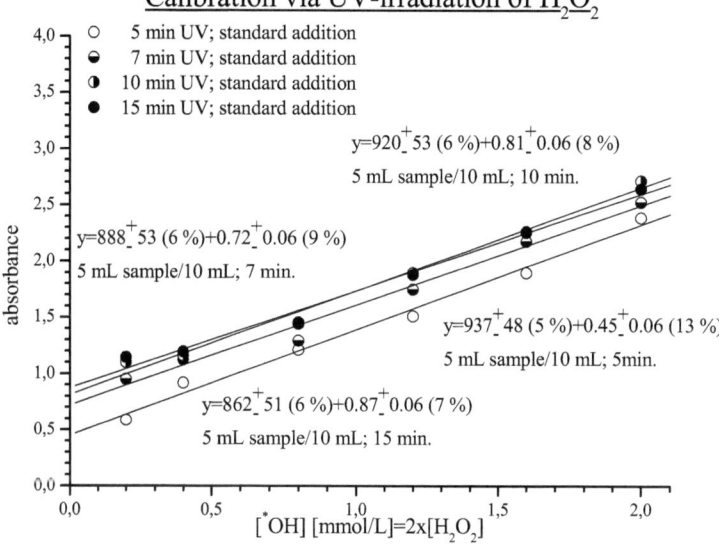

Figure 58 - Calibration of radical trapping with MeOH with standard addition

Standard addition is the better choice when the matrix of the sample may change the analytical sensitivity of the method. Hence it was used to study how the pH of the sample affected the determination of radical concentration. Undiluted H_2O_2 samples, the pH values of which were brought to 1, 3.5 and 9, were irradiated for 10 minutes and their radical concentration determined. The pH values were chosen to cover the working range of values which can be expected in the stripping environment. The values obtained are represented in the calibration curves in Figure 59 and Table XV gives the radical concentrations calculated therefrom.

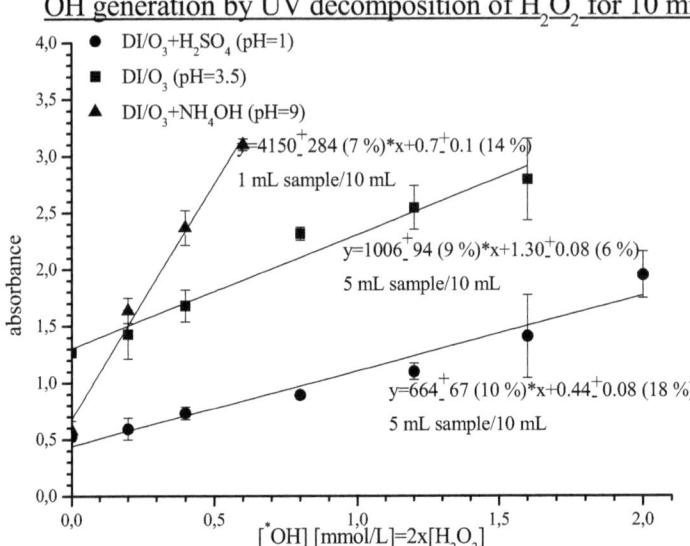

Figure 59 - Radical determination at different pH with MeOH

As expected the results of MeOH trapping show an increase in the radical concentration from pH = 1 to 3.5. Unfortunately this trend could not be observed from pH = 3.5 to pH = 9. It is somewhat odd that the slope increases while the y-offset decreases with the pH. This suggests that a change in pH interferes with the trapping mechanism.

The plots for the 3 pH levels were not parallel and the slopes differed. This called for a reappraisal of the method of direct calibration as a calibration curve would have to be made for every change in the sample matrix. Direct calibration would not then be the simple and quick method it was intended to be and standard addition has to be the method of choice when MeOH is used as the trapping agent.

4.2.1.2. Trapping with DMSO

The conversion of CH_2O to DDL in the presence of an excess of DMSO worked just as well as with MeOH, with calibration curves (Figure 60) and slopes comparable to those obtained with MeOH and free from any influence of excess DMSO.

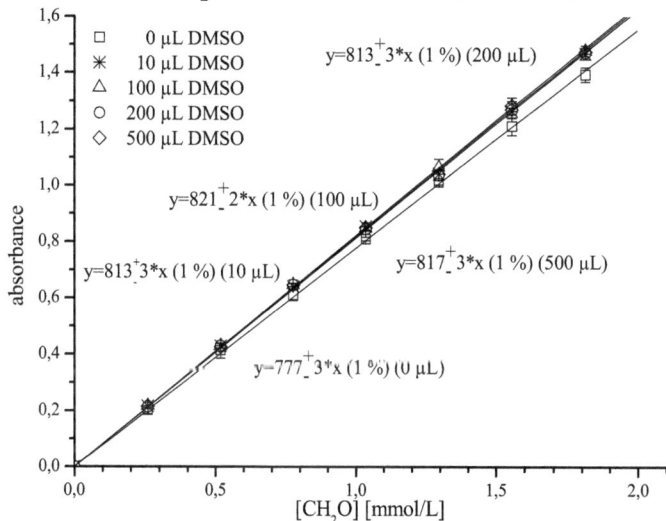

Figure 60 - Calibration of CH_2O conversion to DDL with DMSO in excess

Results and Discussions – Radical determination

The calibration of radical trapping with DMSO using the direct method with both concentrated and diluted solutions (Figure 61) show results that diverge from those obtained with MeOH. The difference lies in the fact that the curve obtained with dilution of the sample is not linear so that this variation of the method has to be ruled out. With concentrated samples the plots are linear and 10 min. of UV irradiation can be taken as the optimum. The slope increases steadily with irradiation time but is disproportionate at 15 min.

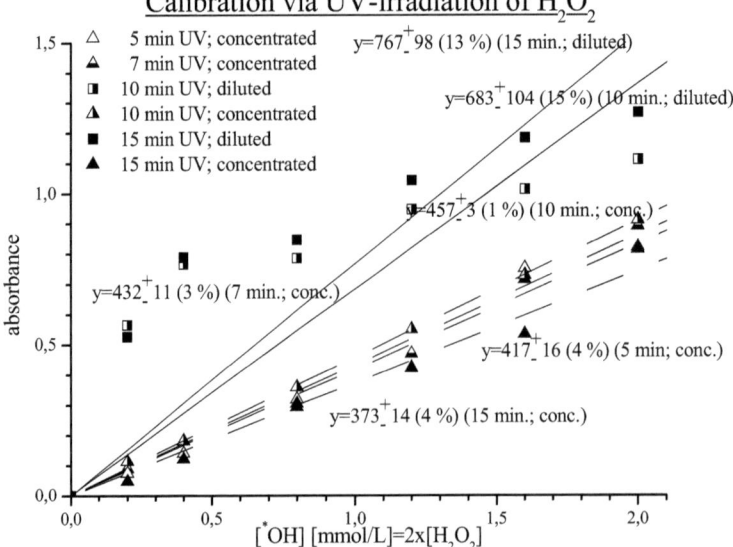

Figure 61 - Calibration of radical trapping with DMSO in diluted & concentrated form

Results and Discussions – Radical determination

The behaviour of the calibration curves for the standard addition with DMSO is quite different from those with MeOH. For DMSO the slope increases with the irradiation time and the y-offset is nearly independent of the pH value. The results for the calculated radical concentrations are presented in Table XIV.

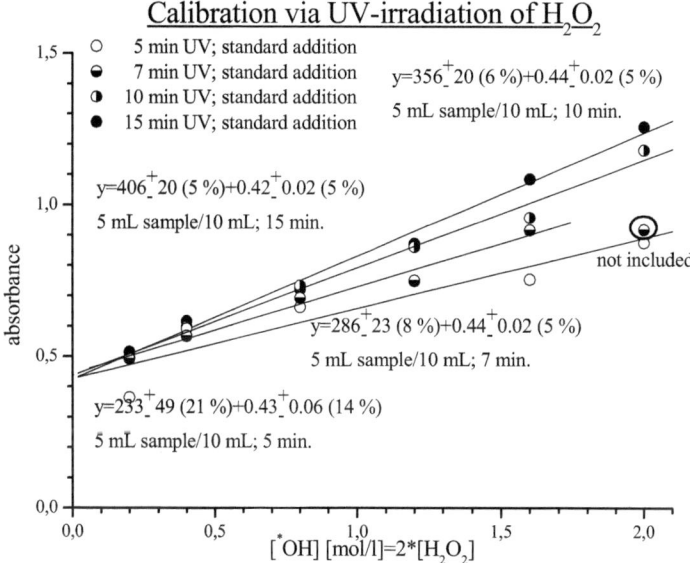

Figure 62 - Calibration of radical trapping with DMSO with standard addition

The application of the standard addition method with DMSO for the same three pH-values as used for MeOH resulted in the calibration curves in Figure 63 and in the calculated radical concentrations in Table XV.

Results and Discussions – Radical determination

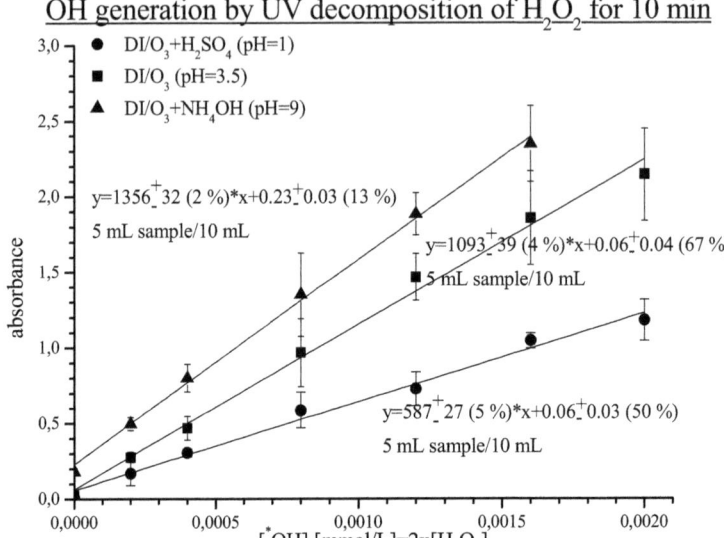

Figure 63 - Radical determination at different pH with DMSO

Table XIV - *OH conc. derived: MeOH vs. DMSO, standard addition, different irradition times

UV-irradiation	[·OH] $\left[\frac{mol}{L}\right]$	
	MeOH (Figure 58)	DMSO (Figure 62)
5	$0.79 \times 10^{-3} - 1.15 \times 10^{-3}$	$2.62 \times 10^{-3} - 5.33 \times 10^{-3}$
7	$1.40 \times 10^{-3} - 1.87 \times 10^{-3}$	$2.72 \times 10^{-3} - 3.50 \times 10^{-3}$
10	$1.54 \times 10^{-3} - 2.01 \times 10^{-3}$	$2.23 \times 10^{-3} - 2.74 \times 10^{-3}$
15	$1.77 \times 10^{-3} - 2.29 \times 10^{-3}$	$1.97 \times 10^{-3} - 2.38 \times 10^{-3}$

Table XV - $^{\bullet}$OH conc. derived: MeOH vs. DMSO by standard addition at different pH values

pH	[·OH] $\left[\dfrac{mol}{L}\right]$	
	MeOH (Figure 59)	DMSO (Figure 63)
1	$0.98 \times 10^{-3} - 1.74 \times 10^{-3}$	$0.1 \times 10^{-3} - 0.32 \times 10^{-3}$
3.5	$2.22 \times 10^{-3} - 3.03 \times 10^{-3}$	$0.04 \times 10^{-3} - 0.19 \times 10^{-3}$
9	$1.35 \times 10^{-3} - 2.07 \times 10^{-3}$	$0.29 \times 10^{-3} - 0.39 \times 10^{-3}$

The slopes of the plots deviate considerably from those obtained with the corresponding MeOH trapping calibration plots (Figure 59). According to the trapping mechanism explained in chapter 2 (2.3.1 Figure 24 and Figure 27) the slopes of the MeOH trapping plot should be twice those of their DMSO counterparts. However, they vary from slightly below the corresponding MeOH plot at pH 3.5 to 3 times that of the MeOH plot at pH 9 (with NH$_4$OH). When the method of standard addition was used on samples of different pH values the radical concentrations derived from MeOH trapping were significantly higher than those obtained with DMSO. One reason for this may be the AOPs mentioned in chapter 2 (2.3.1, Figure 25 and Figure 26).

4.2.2. The combined approach of iodometry and UV/Vis spectroscopy

As the trapping methods with MeOH and DMSO differ significantly in the results they produce, another method, ideally a self-contained method, of determining the radical concentrations was needed to verify the results. In this case this was the method of iodometric titration combined with UV/Vis-spectroscopy. As mentioned in the experimental section the addition of the catalyst $(NH_4)_6Mo_7O_{24}$ should increase the speed of the iodine formation. Although, theoretically, the catalyst should have no influence on the concentration of iodine formed from the oxidising species to be detected, the results indicate the opposite. It is likely that the catalyst also promotes the oxidation of the iodine by atmospheric O_2 and is thus responsible for distorting the results.

Table XVI - Results of radical determination with the iodoemtric-UV/Vis approach

without catalyst				with catalyst			
mL 0.1 M $S_2O_3^{2-}$	mol $S_2O_3^{2-}$	mol $S_2O_3^{2-}$ consumed by O_3	·OH $\left[\dfrac{mol}{L}\right]$	mL 0.1M $S_2O_3^{2-}$	mol $S_2O_3^{2-}$	mol $S_2O_3^{2-}$ consumed by O_3	·OH $\left[\dfrac{mol}{L}\right]$
2.66	2.66×10^{-4}	1.8×10^{-4}	0.86×10^{-3}	2.9	2.9×10^{-4}	1.8×10^{-4}	1.1×10^{-3}
2.50	2.5×10^{-4}	1.8×10^{-4}	0.70×10^{-3}	2.75	2.75×10^{-4}	1.8×10^{-4}	0.95×10^{-3}
2.51	2.51×10^{-4}	1.8×10^{-4}	0.71×10^{-3}	2.81	2.81×10^{-4}	1.8×10^{-4}	1.01×10^{-3}
2.20	2.2×10^{-4}	1.8×10^{-4}	0.40×10^{-3}	/	/	/	/
2.30	2.3×10^{-4}	1.8×10^{-4}	0.50×10^{-3}	/	/	/	/
2.37	2.37×10^{-4}	1.8×10^{-4}	0.57×10^{-3}	/	/	/	/
2.35	2.35×10^{-4}	1.8×10^{-4}	0.55×10^{-3}	/	/	/	/
2.41	2.41×10^{-4}	1.8×10^{-4}	0.61×10^{-3}	2.82	2.82×10^{-4}	1.8×10^{-4}	1.02×10^{-3}
±0.15	$\pm0.15\times10^{-4}$	/	$\pm0.15\times10^{-3}$	±0.08	$\pm0.08\times10^{-4}$	/	$\pm0.07\times10^{-3}$

Results and Discussions – Radical determination

All the methods employed for the radical determination provide reasonable results. Each trapping reaction can be calibrated with good linearity over the whole range required later in practice. In both cases the standard addition is favoured as it allows the investigation in different pH environments. Unfortunately the results of the 2 trapping methods differ substantially.

Iodometric titration in the varied form adopted gave reasonable results for samples in the acidic pH range and were closer to those derived from MeOH trapping. Iodometric titration works in an acidic medium (pH < 2) and its use is therefore limited to samples that are acidic to begin with. Acidifying non-acidic samples would alter their characteristics and, most likely, the radical concentration.

Table XVII - Comparison of trapping results for DI/O_3 at pH=1

method	lower limit [·OH] DI/O_3 pure	upper limit [·OH] DI/O_3 pure
MeOH trapping	0.98×10^{-3}	1.74×10^{-3}
DMSO trapping	0.1×10^{-3}	0.32×10^{-3}
iodometric titration	0.61×10^{-3}	1.02×10^{-3}

4.3. Resist characterization with IR spectroscopy

As mentioned in chapter 2 (2.4) IR spectroscopy places some demands on the samples to be investigated. They need to have polar groups in order to be IR-active and the absorbance needs to be strong enough to be recorded against the background. In the case of layers, as is the case here, this means each layer must be thick enough to allow the interaction with the IR beam. Fortunately the photoresists we are dealing with, especially the unimplanted M91Y and UV26, fulfil all these requirements as they produce detectable, sharp peaks with a high signal-to-noise ratio at any given stage in the stripping process. Thus simple and direct measurements in transmission mode are sufficient there is no need for more demanding setups like directed reflection or ATR. However, this does not apply to the implanted resist DUV248 where, with increasing dopant dosage, the signal becomes weaker, the peaks are less defined and the signal-to-noise ratio is low so that the spectrum of the As 10^{16} implanted DUV248 (Figure 64) does not provide any useful information. This can be easily explained by the formation of a crust during ion implantation which becomes thicker with increasing implant energy and dosage. This crust is assumed to consist of highly cross linked carbon in various non-polar modifications which are not detectable in IR spectra. For these resists Raman spectroscopy was used instead.

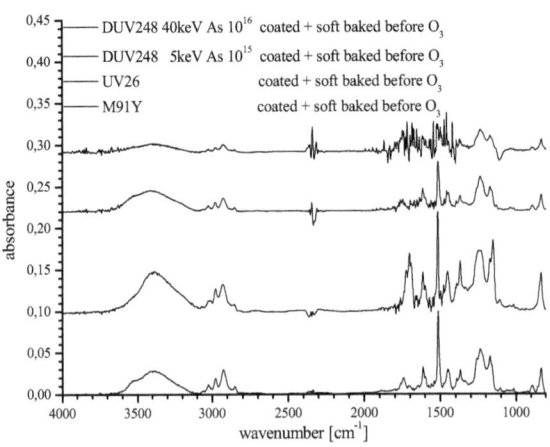

Figure 64 - Comparison of IR spectra of M91Y; UV26; DUV248 as applied

Starting with the spectra of the resist as applied new spectra were recorded after each stage in the process and differential spectra calculated to trace the changes that occurred. In the first stage - the exposure (Figure 65) - the formation of the acid from the PAG cannot be followed with the IR spectrum. Assuming that the PAG in this case were the standard DPI or TPS (Figure 106) this is not astonishing as during their conversion they do not undergo any significant structural changes as the aromatic structures stay intact and the acid generated is not yet able to remove the protection groups.

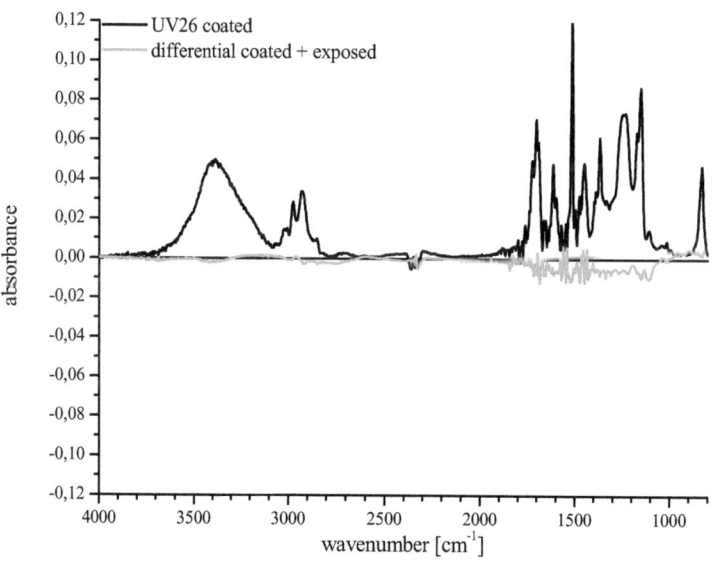

Figure 65 - Resist changes during exposure

Results and Discussions – Resist characterization with IR spectroscopy

The first change visible with IR is the removal of the protection groups during the PEB (Figure 66). During this process an ester group is cleaved. This cleavage is represented by the disappearance of the peak at 1150 cm^{-1}. As there are OH-groups already present belonging to the unprotected styrene parts (more than 50 % unprotected) the additional ones generated by the deprotection nearly carry weight.

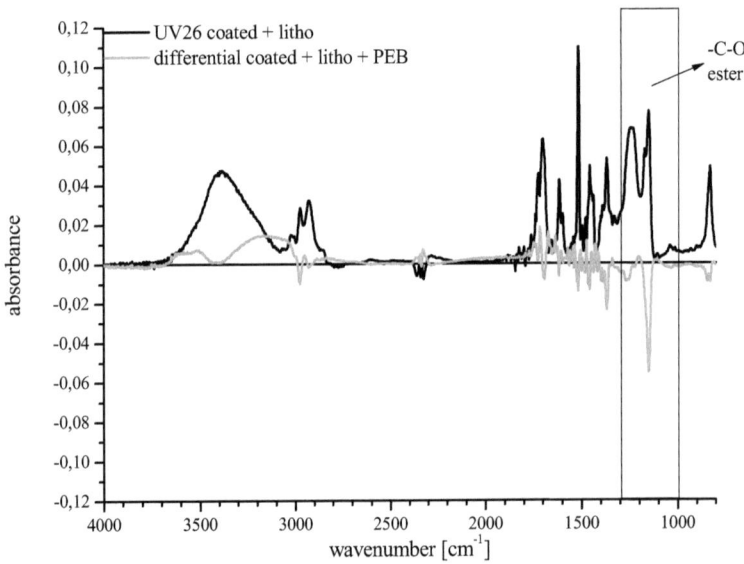

Figure 66 - Resist changes during PEB

The last process stage to be investigated is the plasma etch. As the plasma etch is designed to generate gaseous compounds (often CO_2) the changes to the resist along the way cannot be traced with IR spectroscopy (Figure 67). The loss in resist thickness for these samples during the etch process is only about 145 nm (determined by profilometric measurements) and is not reflected in the general decrease in the size of the peaks.

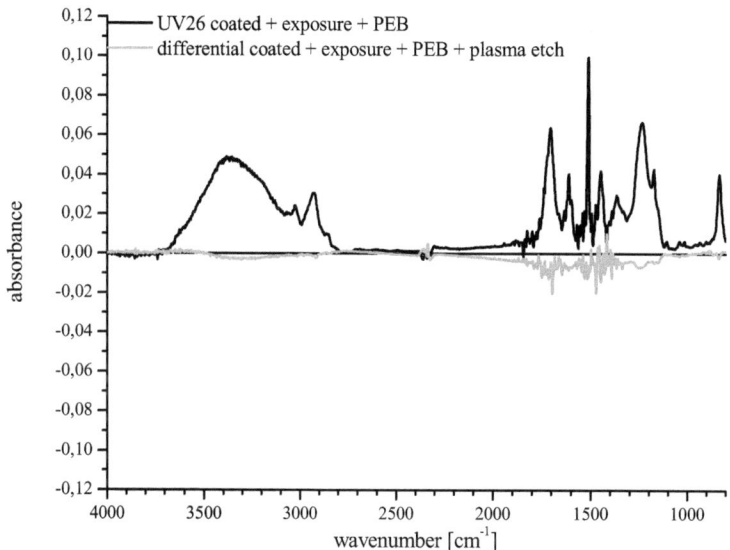

Figure 67 - Resist changes during plasma etch

Results and Discussions – Resist characterization with IR spectroscopy

After recording the spectra of wafer surfaces during the various stages of resist treatment, their spectra during resist stripping with ozone were also recorded.

The spectra in (Figure 68 - Figure 72) represent the resist at various process stages whereby the black curve always represents the spectra before partial O_3-treatment and the red one the differential spectra after 5 min of O_3 treatment.

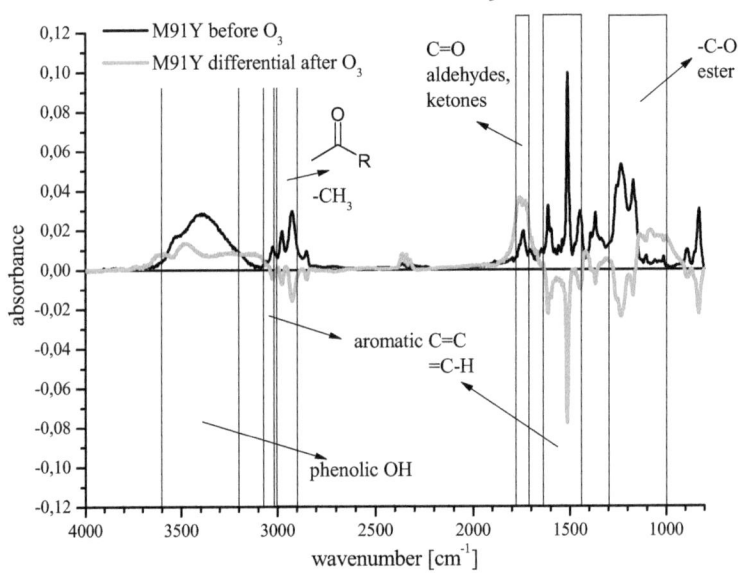

Figure 68 - IR spectrum of JSR KrF M91Y before and after ozone treatment

Results and Discussions – Resist characterization with IR spectroscopy

As all resists are of the PHS-type their spectra are nearly identical. Therefore only the spectrum of the coated + soft baked UV26 will be discussed.

With the knowledge of the resist composition as given for Rohm & Haas UV26 the peaks of the IR spectrum can be assigned to the functional groups of the resist (Figure 69). Beginning with the low wavenumbers the peaks can be assigned as followed:

1000 - 1300 $\tilde{\nu}$: -C-O-C in esters
1450 - 1650 $\tilde{\nu}$: C=C; =C-H in aromatic systems
1700 - 1800 $\tilde{\nu}$: C=O in aldehydes or ketones
2900 - 3000 $\tilde{\nu}$: -CH$_3$ in tert. butyl groups
3000 - 3100 $\tilde{\nu}$: C=C; =C-H in aromatic systems
3200 - 3600 $\tilde{\nu}$: -OH in phenols (wide range for -OH in various compounds)

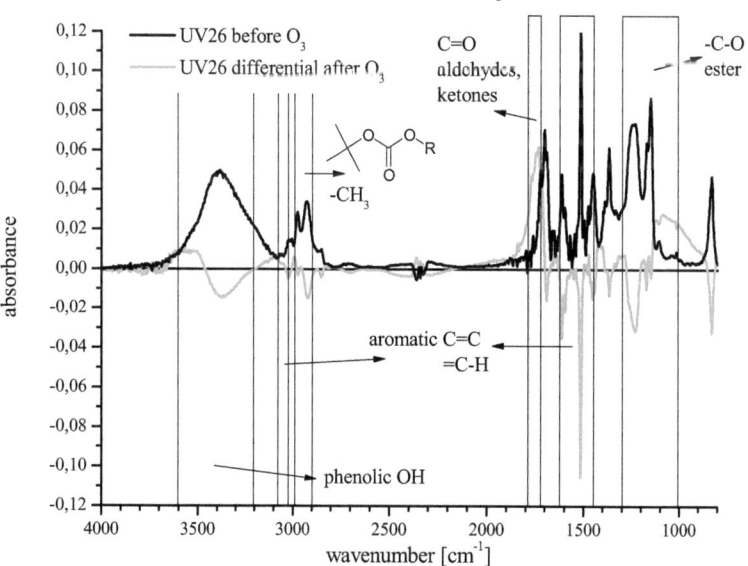

Figure 69 - IR spectrum of Rohm&Haas UV26 before and after ozone treatment

Results and Discussions – Resist characterization with IR spectroscopy

The differential spectra support the assumption that the decomposition mechanism for the resists is a classical ozonolysis. This is indicated by the change in the peaks as follows:

a decrease at 1000 - 1300 $\tilde{\nu}$ $\Rightarrow$ decomposition of C-O-C bond of the ester group

a strong decrease at 1500 $\tilde{\nu}$ $\Rightarrow$ decomposition of C=C bonds of the aromatic group

an increase at 1700 - 1800 $\tilde{\nu}$ $\Rightarrow$ formation of the C=O group of ketones or aldehydes

a decrease at 2900 - 3000 $\tilde{\nu}$ $\Rightarrow$ removal and conversion of the t-BOC protection group

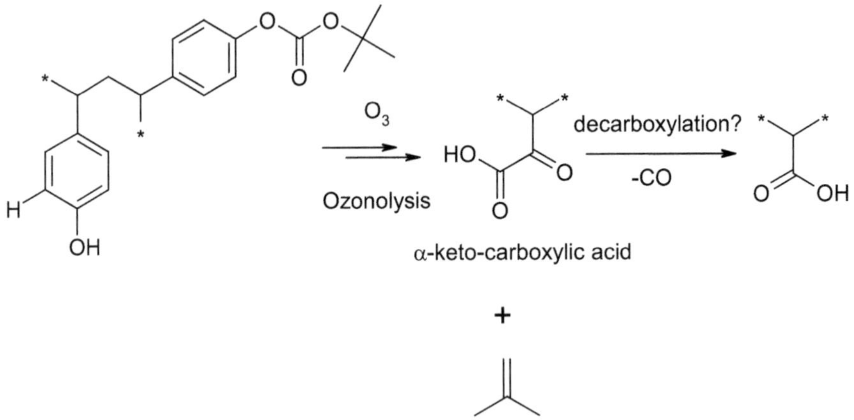

Figure 70 - Decomposition of the Rohm&Haas UV26 resist

From the stripping results discussed later on it appears that the removal of the protection group is the first step, followed by the ozonolysis of the aromatic bonds. A possible further decomposition of the α-keto carboxylic acid by decarboxylation in acidic media (H_2SO_4 soln.) [25] is a well know reaction and might also occur here leading to a conversion of the α-keto acid to the homologous shorter simple carboxylic acid.

Results and Discussions – Resist characterization with IR spectroscopy

As already mentioned the spectra of the implanted resists always show peaks with intensities lower than those of unimplanted resists. In the low-energy low-dose As implanted resist a large part of the resist remains unmodified and is therefore detectable in IR spectra. As DUV248 has the same structure as M91Y the spectra are the same.

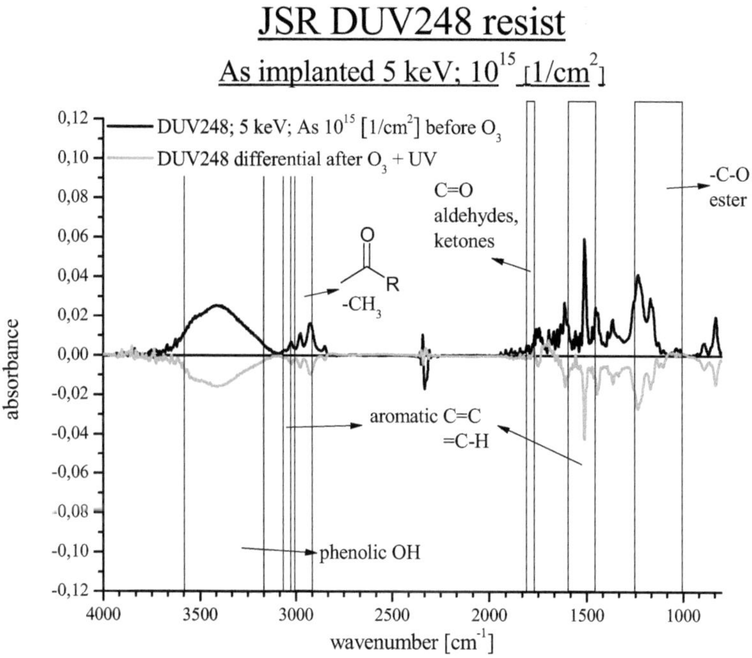

Figure 71 - IR spectrum of JSR DUV248 5 keV before and after ozone treatment

In the high-energy high-dose As implanted resist the modification affects a larger part of the resist layer making it IR inactive. This effect combined with a lack of response to the ozone treatment results in changes not being detectable in the highly implanted resist.

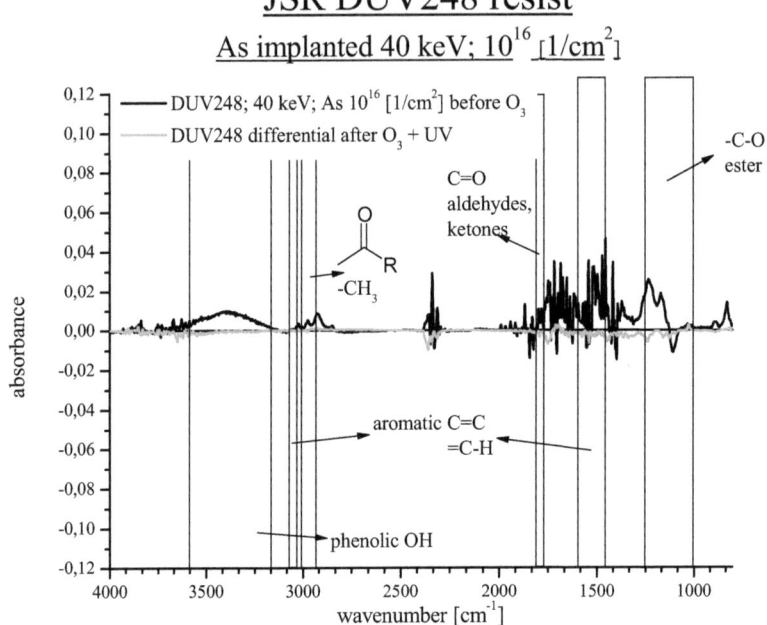

Figure 72 - IR spectrum of JSR DUV248 40 keV before and after ozone treatment

4.5. Resist characterization with Raman microscopy

As explained in chapter 2 (2.5) and practically illustrated by the IR spectra for the implanted resists the ion implant process changes the resist structures towards non-polar ones not detectable by IR. Raman spectroscopy offered a means to complement the IR spectra.
As illustrated by Figure 73 Raman spectroscopy delivers detectable peaks over the whole working range of 150 - 4000 $\tilde{\upsilon}$.

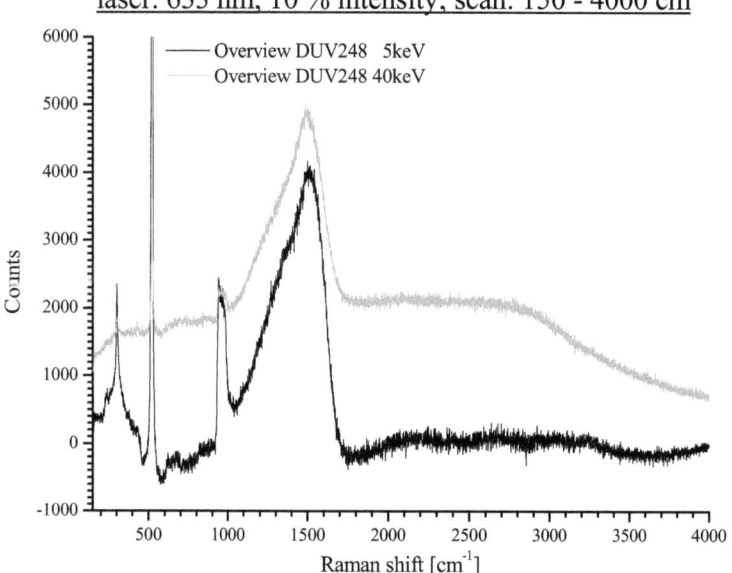

Figure 73 - Raman overview spectra of DUV248 5 and 40 keV

A comparison of each resist spectrum with the reference spectrum measured on a resist-free wafer portion of the wafer (Figure 74, Figure 75) shows that all peaks below 1000 $\tilde{\upsilon}$ belong to the Si-substrate and only the wide peak at 1100 - 1700 $\tilde{\upsilon}$ comes from the resist.

Results and Discussions – Resist characterization with Raman microscopy

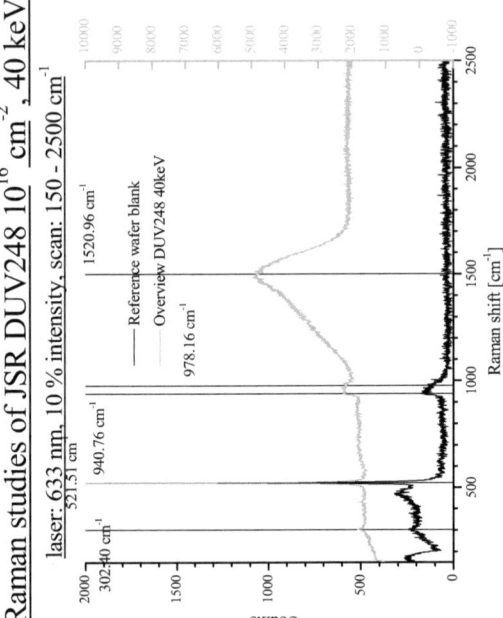

Figure 74 - Raman spetrum of DUV248 As 10^{15} 5 keV vs. blank Si

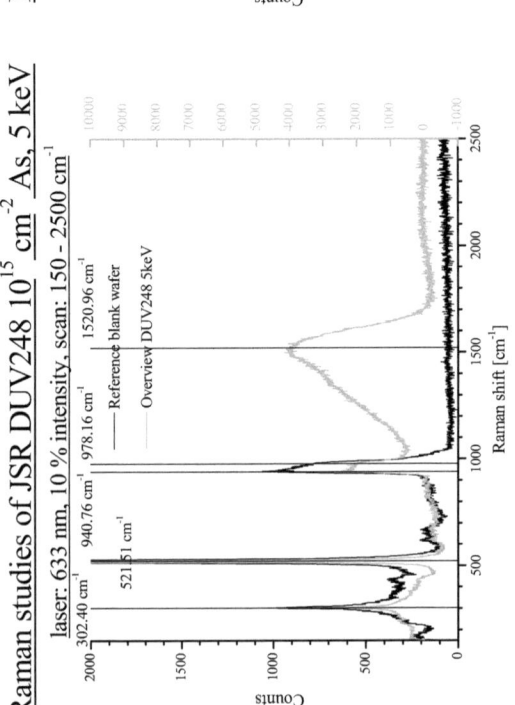

Figure 75 - Raman spetrum of DUV248 As 10^{16} 40 keV vs. blank Si

Results and Discussions – Resist characterization with Raman microscopy

Once Raman spectroscopy had proven its applicability for the study of implanted resists in general, a further variation, Raman microscopy, was used to try and distinguish between unmodified (bulk) resist and the crusted areas (top crust) of the resist by depth profiling. Spectra were recorded of layers focussed on in progressive steps from the top crust down to deep below the resist. Not only was there no change in the spectra but in all cases the appearance of peaks at 940 cm^{-1} and 978 cm^{-1} attributable to the silicon wafer substrate suggest that the resist layer is too thin for such depth profiling.

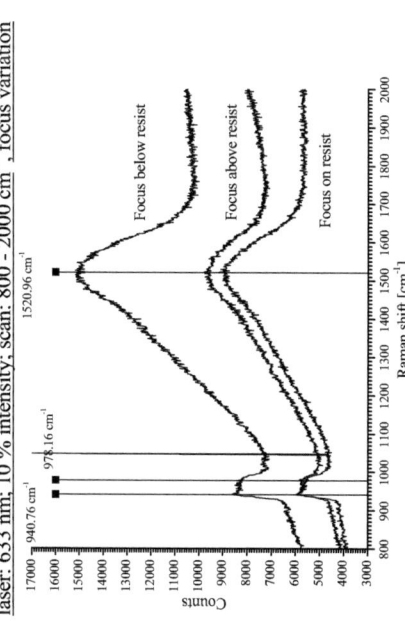

Figure 76 - Raman spectra of DUV248 5 keV with different depths of focus

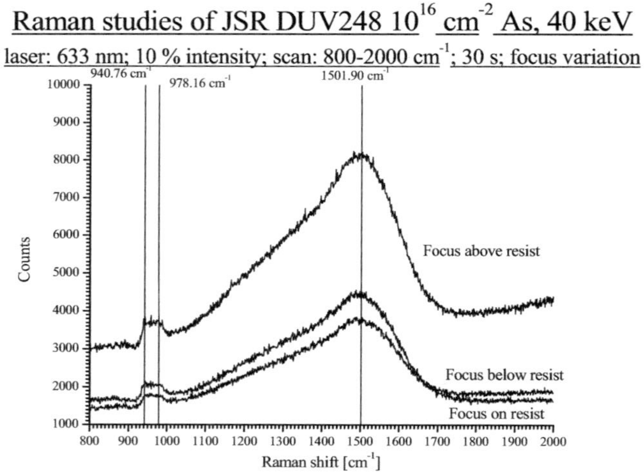

Figure 77 - Raman spectra of DUV248 40 keV with different depths of focus

Results and Discussions – Resist characterization with Raman microscopy

As the unmodified resist should be Raman inactive the broad peak extending from 1030 cm^{-1} to 1730 cm^{-1} most likely represents the crust and consists of more than a single compound or phase. Presuming that the crust is highly cross linked and consists mainly of carbon, peak positions of the most likely carbon modifications have been marked in the spectra and suggest that the crust is made up of graphite in various modifications but without a structural order. It is interesting to note that the two shoulders of the peak correspond to the reference peaks but there is no match for the main peak at 1500 cm^{-1}.

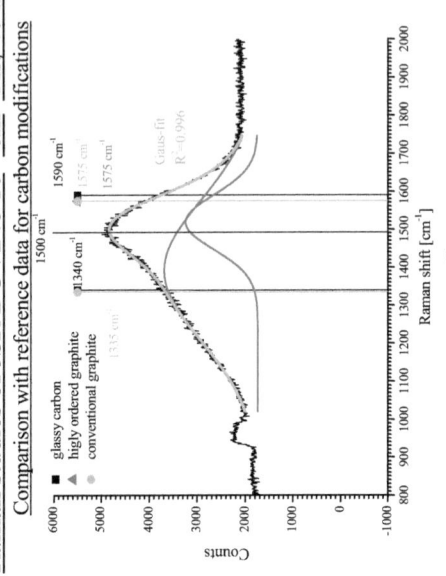

Figure 78 - DUV248 As 10^{15} 5 keV peak allocation

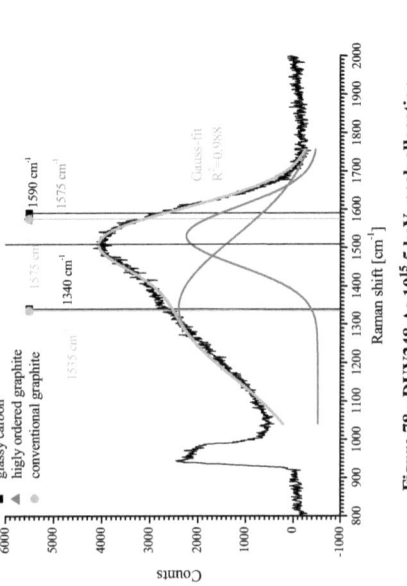

Figure 79 - DUV248 As 10^{16} 40 keV peak allocation

The two peaks actually overlap as can be seen in the Gaussian curve fit of the two reference peaks (green) at 1030 cm^{-1} and 1575 cm^{-1} which results in a reference curve (red) that accurately matches the experimental peak at 1500 cm^{-1} and 1510 cm^{-1}. The broadness of the peaks can be explained by the amorphous character and the possible mixture of different modifications and phases within the crust since highly ordered crystalline structures of a single modification would produce sharp peaks.

4.6. Resist stripping

As the aim of this work is to understand the chemistry of DUV resist stripping with ozone and to develop methods and find suitable additives that will improve the stripping at all stages of the process, a crucial step is the study of the influence of additives on the stripping efficiency of ozone. For this purpose different additives were studied as described in chapter 3 (3.5). The results for a complete resist strip of the Rohm&Haas UV26 resist sorted according to the pH value of the stripping solution and the various stages in the process are presented in Figure 80 and Figure 81.

Increasing the temperature to 50 °C was the one factor that reduced the stripping time with all variations of the stripping solution and every stage of the process.

In all cases there was a clear reduction of the stripping time from step 1 (coated + soft baked) towards step 4 (partially plasma etched). This tendency can be explained by the structural changes the resist undergoes during these steps:

step 1: resist with protection groups ⇒ insoluble in developer

step 2: exposure ⇒ generation of the acid from the PAG, PG still on

step 3: PEB ⇒ reaction of the discharged acid and elimination of the protection group ⇒ strongly increased solubility, higher electron density in the arenes (Figure 84)

step 4: partial plasma etch ⇒ resist thinned down by the etch, but formation of etch residues and crust the reason for no further decrease of the stripping time

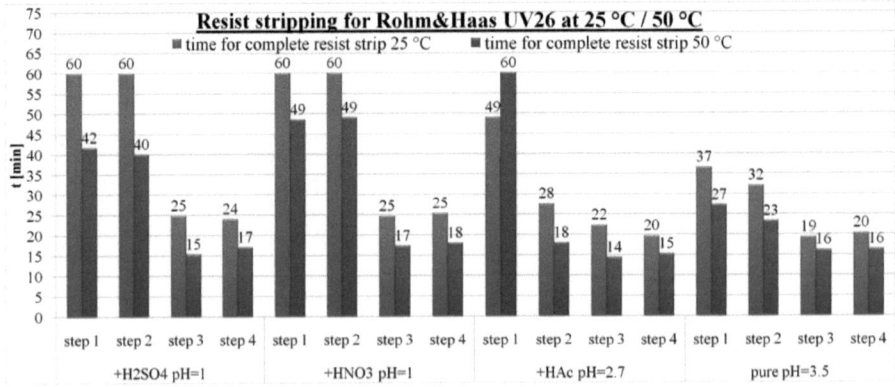

Figure 80 - pH effect on stripping efficiency of DI/O$_3$ on UV26 at 25 °C/50 ° (by pH)

Results and Discussions – Resist stripping

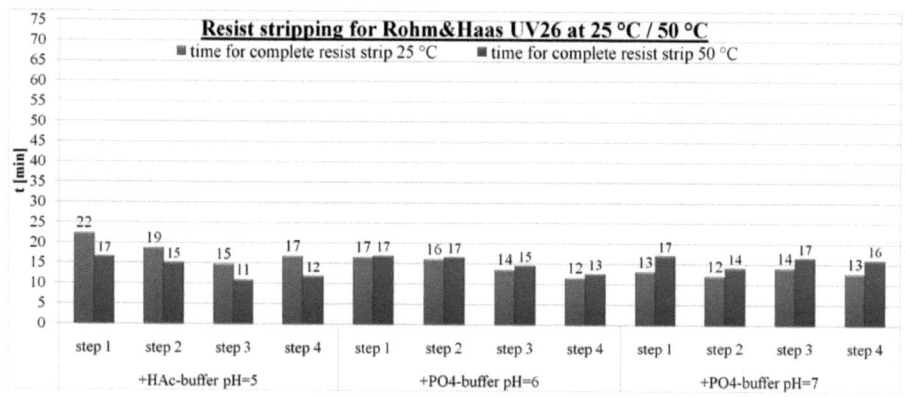

Figure 81 - pH effect on stripping efficiency of DI/O$_3$ on UV26 at 25 °C/50 ° (by pH)

In Figure 82 and Figure 83 the results have been rearranged to give a better picture of the stripping efficiency in relation to pH. For all process steps it is obvious that an increase of the pH from acidic (pH = 1) towards neutral (pH = 7) reduces the time needed for a complete strip to less than half. This effect is strongest for step 1 and 2 and weak for step 3 and 4. Surprisingly using HNO$_3$ as an additional oxidiser gave the worst stripping result, significantly worse than with H$_2$SO$_4$ at pH = 1. So only increasing the oxidising power is not the key to faster resist stripping as the best removal corresponds to the lowest oxidising potential E (Table XVIII). Therefore it is more important to have optimum conditions that support the mechanism of oxidation than oxidising power alone.

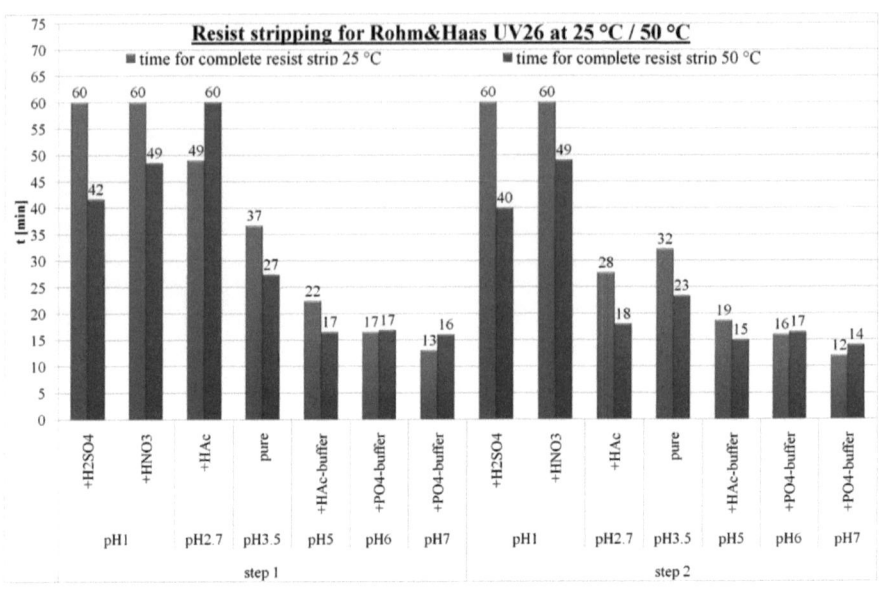

Figure 82 - pH effect on stripping efficiency of DI/O₃ on UV26 at 25 °C/50 ° (by step)

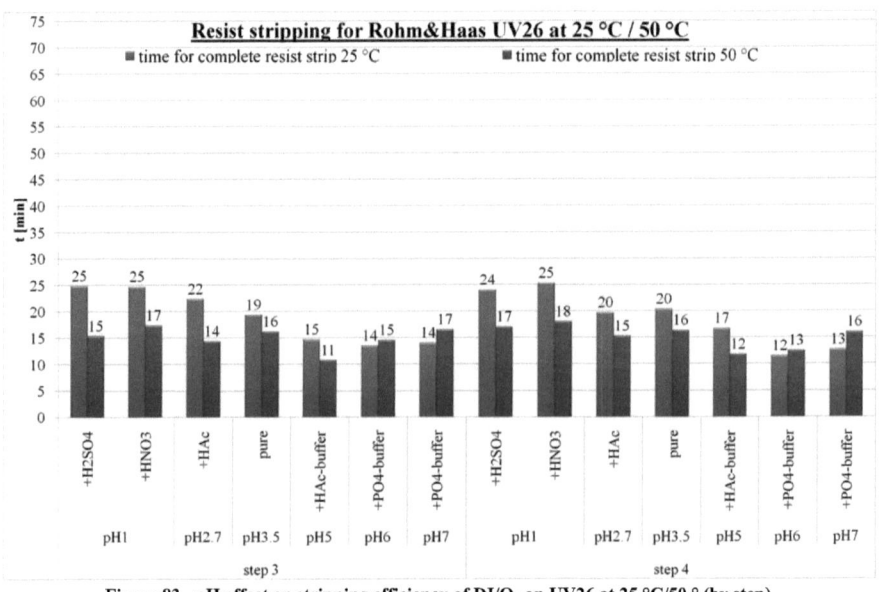

Figure 83 - pH effect on stripping efficiency of DI/O₃ on UV26 at 25 °C/50 ° (by step)

Table XVIII - Oxidising potentials (determined by potentiometric measurements)

	+ H_2SO_4 pH=1	+ HNO_3 pH=1	pure DIW/O_3 pH=3.5	+ HAc-buffer pH=5	+ $H_3PO_4^-$ buffer pH=7
oxidising potential E [V]	1.390	1.396	1.263	1.193	1.066

Regarding the oxidising power it is reduced at higher pH-values.

The general tendency of faster stripping at higher pH levels in a certain range can be explained by the presence of the OH-function of the phenol. As an aromatic substituent oxygen has a negative inductive effect (minus i-effect) and only a positive mesomeric effect (plus m-effect), hence its presence induces the withdrawal of electrons from the aromatic core and therefore causes a deactivation of electrophilic attacks such as ozonolysis. But the phenolic OH-group is more acidic than aliphatic alcohol functional groups (phenol: pK_a = 10, EtOH: pK_a = 15.9). The reason for this is the possibility of delocalisation of the negative charge within the aromatic system. So by increasing the pH-value it is possible to deprotonate the phenolic OH-group and thereby increase the electron density within the aromatic core (Figure 84) making it more reactive towards electrophilic attacks as desired for ozonolysis. As far as this effect is concerned, the higher the pH the better the ozonolysis as even at pH = pK_a = 10 only 50 % of the phenolic hydroxyl groups are deprotonated. But there is a second factor to keep in mind, namely, the amount of avaiable molecular O_3 necessary for ozonolysis which decreases sharply at higher pH values ($\tau_{1/2}$ at pH = 9 < 5 s). So an optimum pH value has to be found which seems to be around pH = 5 - 6.

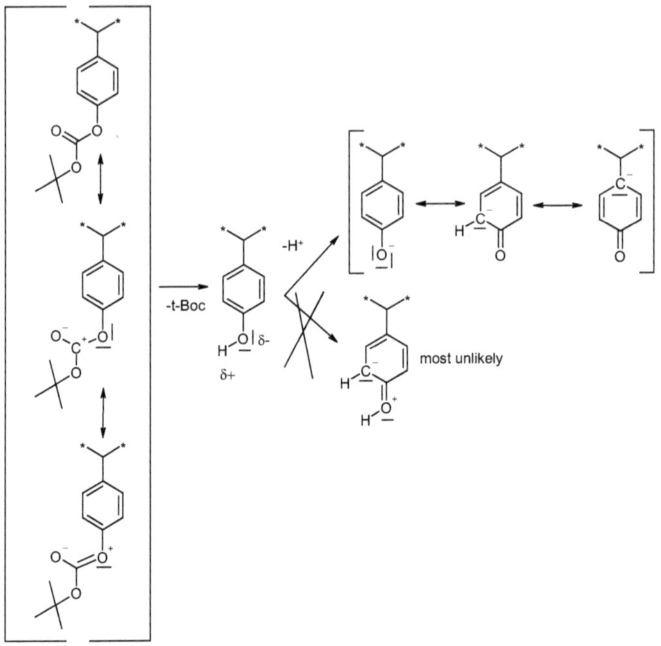

Results and Discussions – Resist stripping

Figure 84 - Phenol deprotonation

The significantly higher effect of increased pH-values (pH=1 → 2.7 or 3.5) for step 1 and step 2 is quite surprising as the PG is base resistant and only to be removed by acids and as mentioned earlier the resist removal within the PEB (step 3) is the reason for the much faster resist stripping during the steps 3 and 4.

Rohm&Haas UV26 and JSR M91Y (Figure 85 - Figure 88) have the same general trend but UV26 was more difficult to remove in steps 1 and 2. In steps 3 and 4 it was not different from JSR M91Y in its behaviour. As both resists are of the PHS type their difference lies in their protection groups (PG) and different levels of protection in step 1 and 2 where the PG is still present. Once the PG is removed in the PEB in step 3 this difference no longer exists.

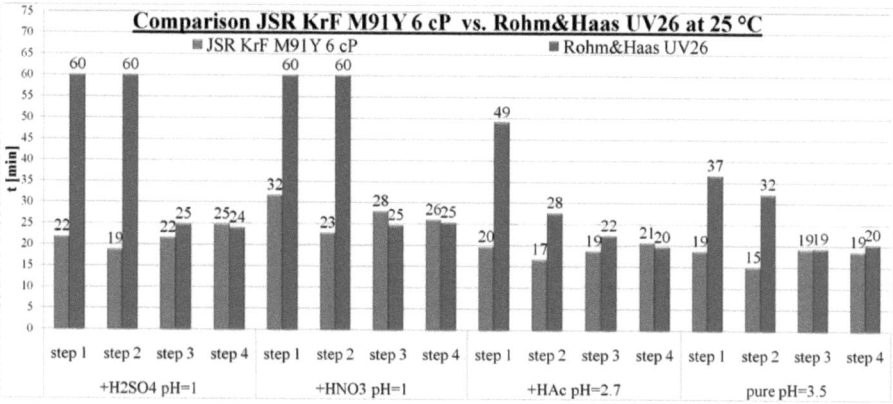

Figure 85 - pH effect on stripping efficiency of DI/O_3 on M91Y/UV26 at 25 °C (by pH)

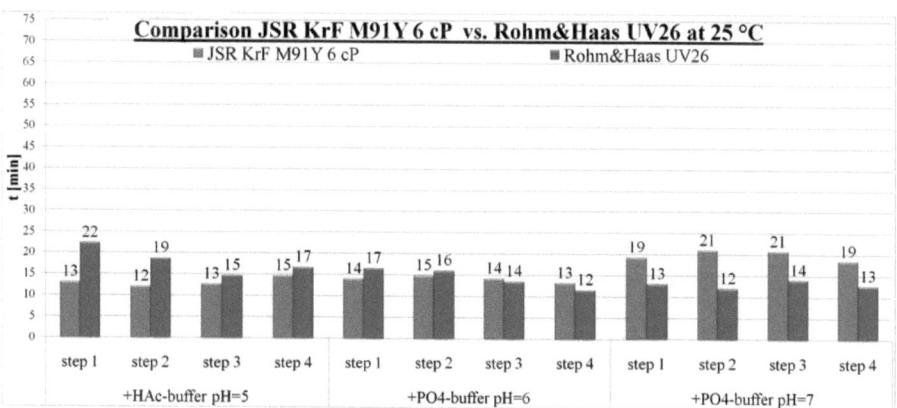

Figure 86 - pH effect on stripping efficiency of DI/O_3 on M91Y/UV26 at 25 °C (by pH)

Results and Discussions – Resist stripping

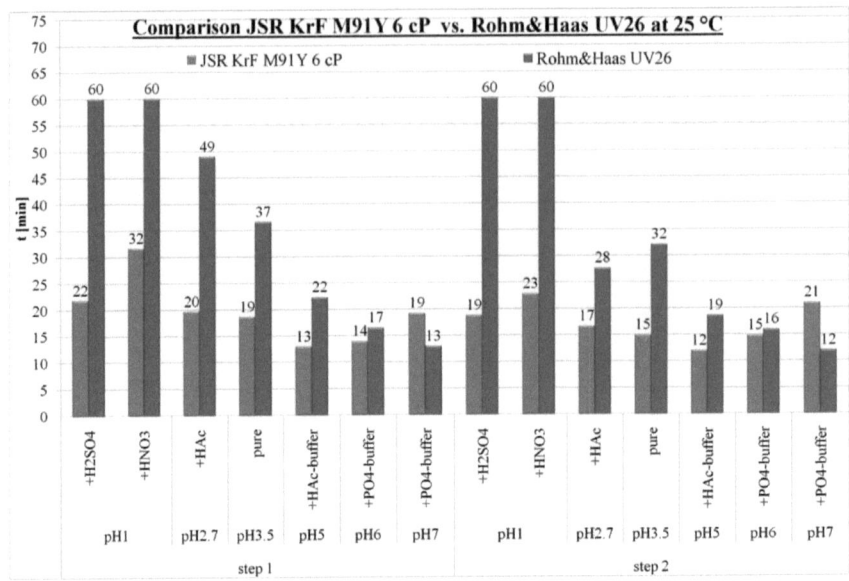

Figure 87 - pH effect on stripping efficiency of DI/O_3 on M91Y/UV26 at 25 °C (by step)

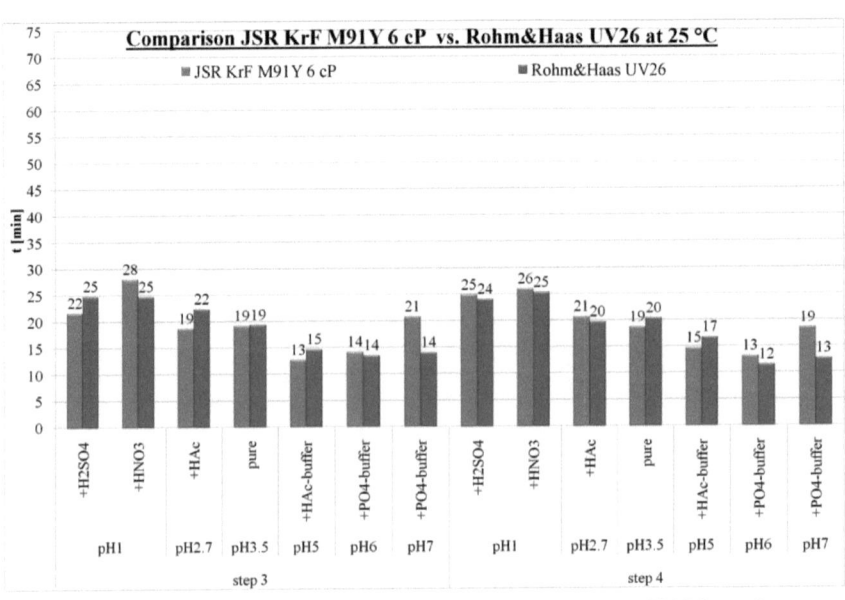

Figure 88 - pH effect on stripping efficiency of DI/O_3 on M91Y/UV26 at 25 °C (by step)

Results and Discussions – Resist stripping

One aspect at least as important as the total time needed for complete resist strip is the behaviour of the resist at various process stages during the strip. This is as an additional indicator of the processes undergone during the strip. To observe this behaviour the resist thickness was measured over time with a profilometer starting with the initial thickness of the resist. The results for the UV26 at 25 °C (Figure 89) show that, as expected, the initial thickness for the steps 1 and 2 is the same and for step 4 lower as a result of the preceding partial plasma strip. Surprisingly the initial thickness for step 3 (after the release of the PAG) is much lower than that in step1 and 2 but equal to the thickness in step 4.

In contrast to the expected constant thinning, the resist seems to require an initiation time which is longer for step 1 and 2 with the PG and shorter for step 3 and 4 without the PG, with an increased thinning rate towards the end.

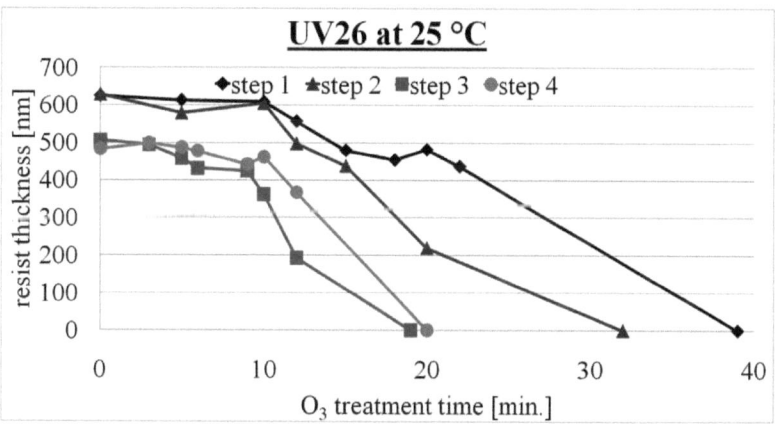

Figure 89 - UV26 resist thinning by DI/O$_3$ at 25 °C

More detailed studies of this process by a surface scan of the resist with the profilometer (Figure 90, Figure 91) indicate a surface roughening at the beginning of the resist strip with the formation of pits within the resist surrounded by walls. With the optical microscope (Figure 92) these pits in the resist can be seen as round spots, were gas bubbles formed on the surface during stripping.

Results and Discussions – Resist stripping

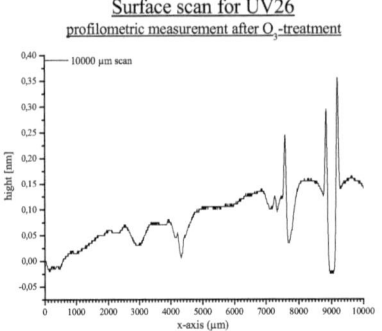

Figure 90 - Surface scan 10000 µm

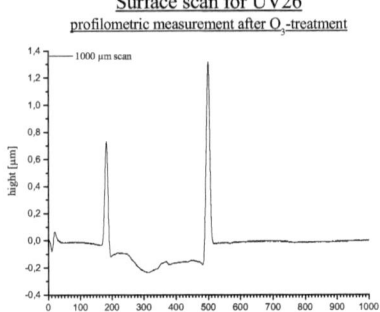

Figure 91 - Surface scan 1000 µm

From these pits the DI/O$_3$ starts to penetrate the resist accompanied by a swelling of the same and followed by thinning of the layer (Figure 93). Finally the thin remaining resist delaminates, coming off in flakes.

Results and Discussions – Resist stripping

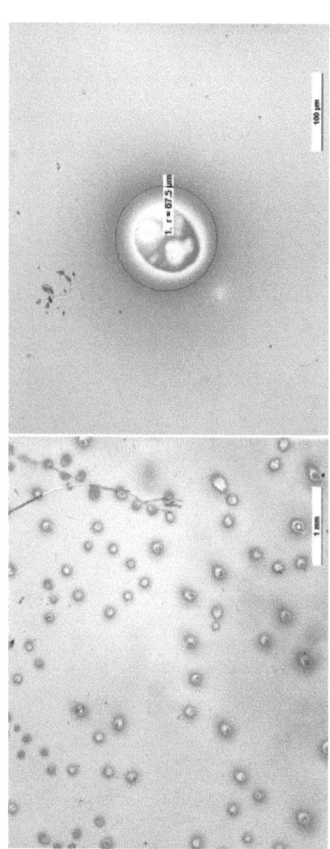

Figure 92 - UV26 DI/O$_3$ for 5 min

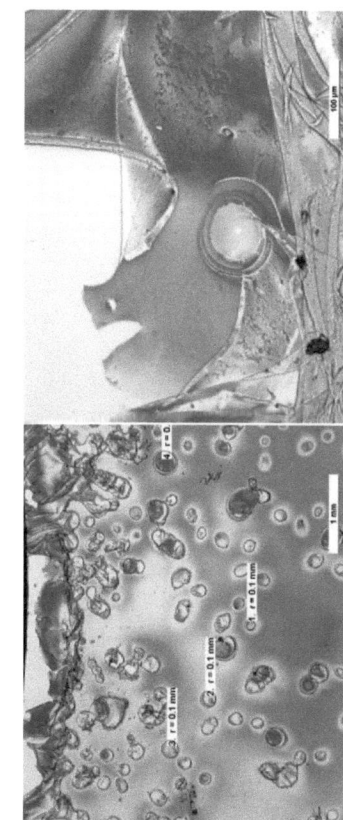

Figure 93 - UV26 DI/O$_3$ for 10 min

Results and Discussions – Resist stripping

As had been feared, neither DI/O$_3$ alone nor DI/O$_3$ in any of its variations as applied to unimplanted resists worked with the implanted resists. In IMEC [43] attempts were made to remove the resist, first with DI/O$_3$ alone at 90 °C. Even after one hour of immersion this had no effect on the low-energy 5 keV $10^{15}\left[\dfrac{1}{cm^2}\right]$ As implanted resist (Figure 94) not to mention the high-energy high-dose 40 keV $10^{16}\left[\dfrac{1}{cm^2}\right]$ As resist. Attempts to remove the resists in an acidic medium (pH = 1) at 90 °C also showed little success.

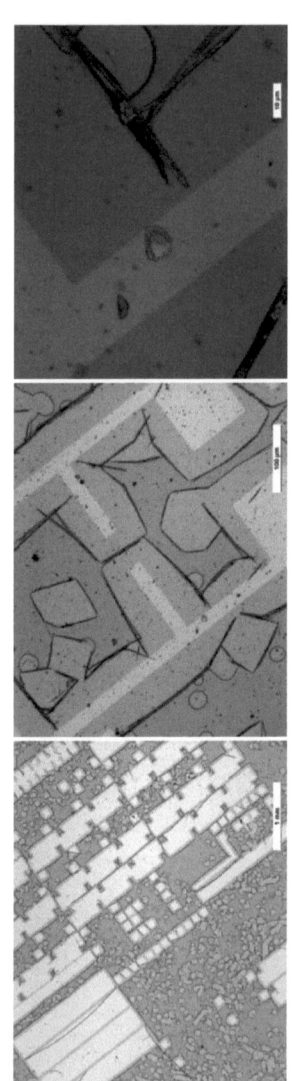

Figure 94 - (As 10^{15} cm^{-2}; 5 keV) – DI/O$_3$ pH=3.5; beaker; 90 °C; 1 h

Results and Discussions – Resist stripping

The next logical step, stripping with the combination of the parameters that worked best for the unimplanted resists (pH = 5 with HAc or pH = 6 with phosphate buffer + high temp. 90 °C) failed to remove either of the implanted resists after one hour.

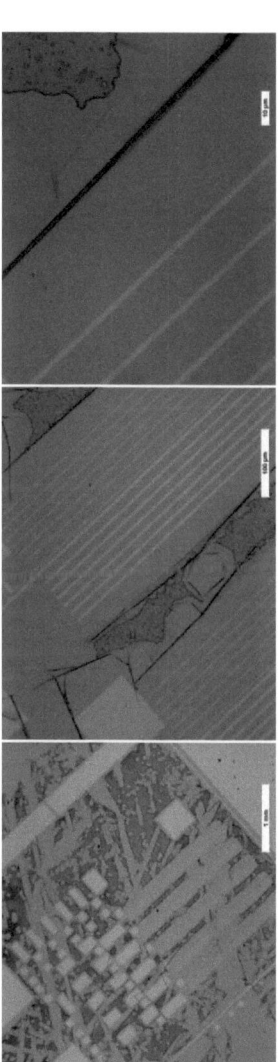

Figure 95 - (As 10^{15} cm^{-2}; 5 keV) – DI/O$_3$ + HAc-buffer pH=5; beaker; 90 °C; 1 h

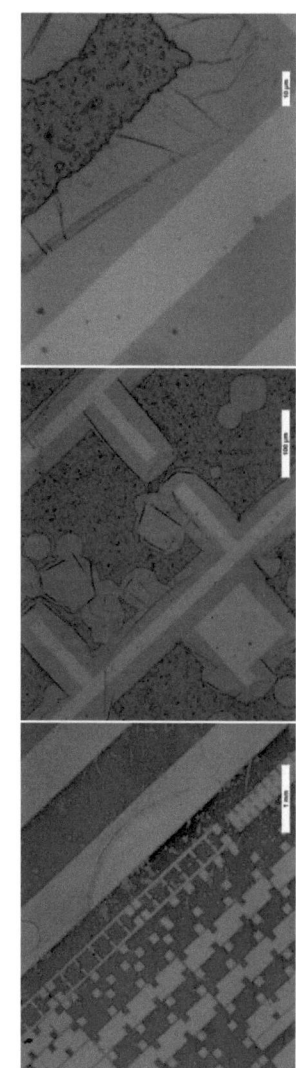

Figure 96 - (As 10^{15} cm^{-2}; 5 keV) – DI/O$_3$ + PO$_4^{3-}$-buffer pH=6; beaker; 90 °C; 1 h

Results and Discussions – Resist stripping

The breakthrough came when instead of the beaker setup (Figure 43 chapter 3) the dispenser setup (Figure 47 in chapter 3) was used, allowing the sample to be irradiated with UV light while the DI/O$_3$ was being dispensed to induce radical generation. With this setup it was possible to strip the 5 keV $10^{15} \left[\frac{1}{cm^2}\right]$ As implanted resist within 10 min at RT (Figure 97).

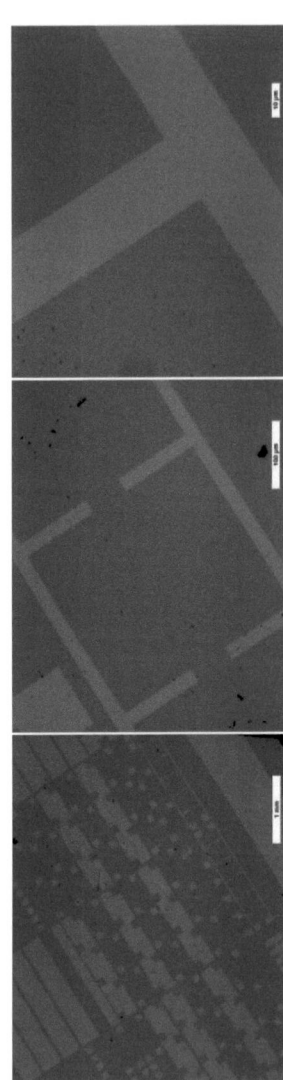

Figure 97 - (As 10^{15} cm^{-2}; 5 keV) – DI/O$_3$ flow + UV; 10 min

The experiment was repeated twice, leaving out the UV light in one (Figure 98) and the O$_3$ in the other (Figure 99). In neither of the experiments was the resist stripped. From this it follows that stripping was effected by the generation of radicals from DI/O$_3$ under UV irradiation.

Results and Discussions – Resist stripping

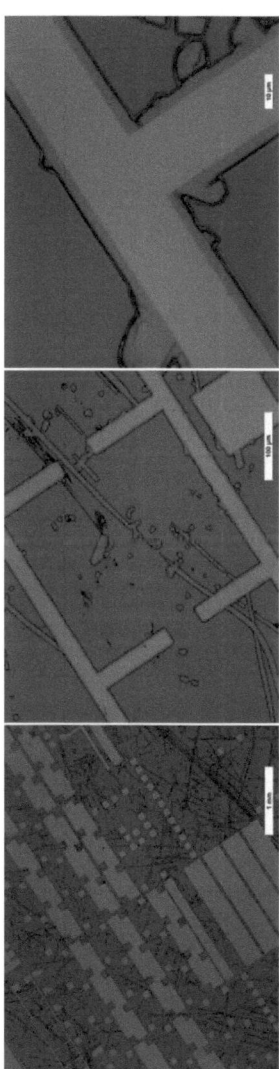

Figure 98 - (As 10^{15} cm^{-2}; 5 keV) – DI/O$_3$ flow; 10 min

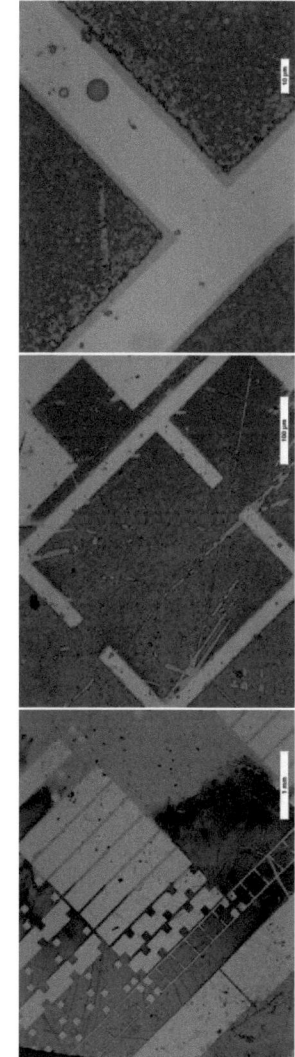

Figure 99 - (As 10^{15} cm^{-2}; 5 keV) – DI flow + UV; 10 min

Results and Discussions – Resist stripping

Although this method worked very well on the low-dose low-energy implanted resist it was not possible to transfer it to the high-energy high-dose 40 keV $10^{16} \left[\frac{1}{cm^2}\right]$ As implanted resist. Even after 1 h of this treatment the resist stripping was rather poor (Figure 100).

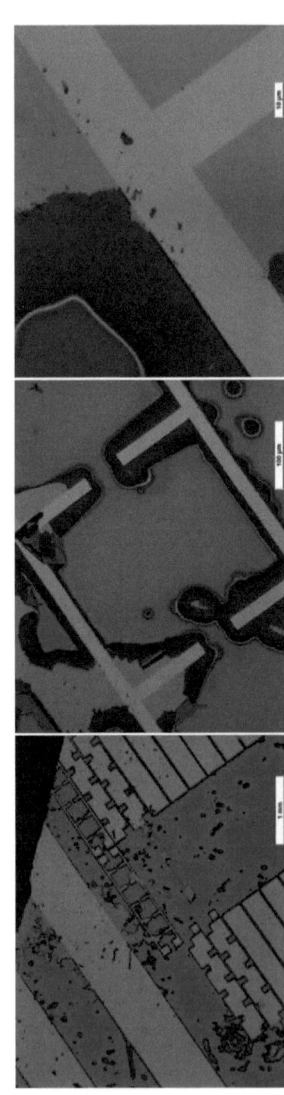

Figure 100 - (As 10^{16} cm^{-2}; 40 keV) – DI/O$_3$ flow + UV; 1 h

Results and Discussions – Resist stripping

Even the phosphate buffered and the HAc buffered DI/O$_3$ (both of which worked best with unimplanted resists) in combination with UV irradiation showed poor results even after one hour of stripping (Figure 132, Figure 133) although the HAc buffered DI/O$_3$ was slightly better than the phosphate buffered solution. DI/O$_3$ brought to pH 7 and 12 using the additives in Table XI gave better results but not a complete strip. At pH 13.5 with KOH as the additive the stripping was considerably better. A successful complete strip was finally obtained for the 5 keV 10^{15} sample with a three minute treatment with DI + KOH alone without UV irradiation or O$_3$ (Figure 101).

Figure 101 - (As 10^{15} cm^{-2}; 5 keV) – KOH pH=13.5 flow; 3 min

Results and Discussions – Resist stripping

This short, simple and very effective method of stripping low-energy low-dose As implanted resists did not work with high-energy high-dose As implanted resists. The highly alkaline medium etched and damaged the Si wafer surface in the one hour necessary for a complete strip (Figure 102).

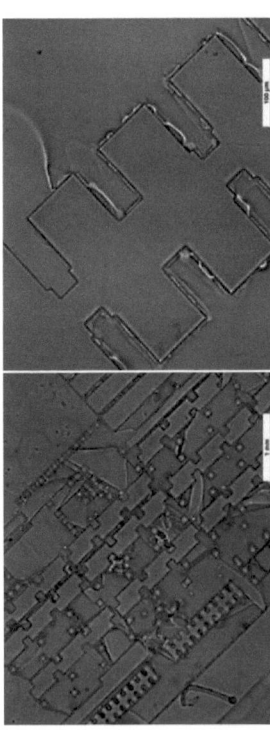

Figure 102 - (As 10^{16} cm^{-2}; 40 keV) – KOH pH=13.5 flow; 1 h

Results and Discussions – Resist stripping

The use of pyrrolidine in H_2O (1:1) which had a pH of 13.12 gave promising results after 45 minutes of stripping with a nearly complete resist removal without damage (Figure 103). However, pyrrolidine is not a commonly used compound in the semi-conductor industry and it is doubtful if it will be accepted. NH_4OH on the other hand was a less complicated way to bring the pH of DI/O_3 to 13 and the results in combination with UV irradiation and DI/O_3 were reasonable.

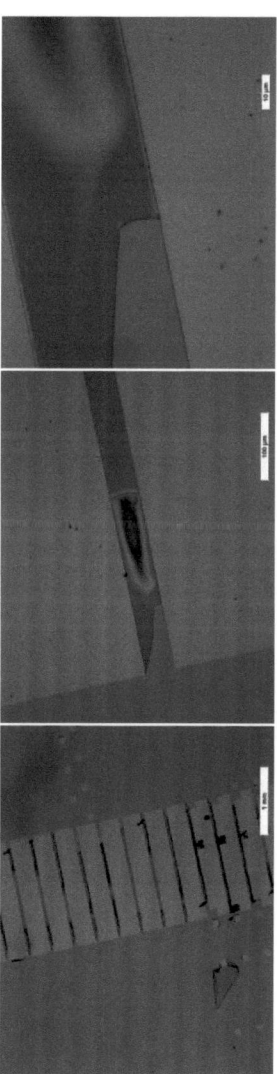

Figure 103 - (As 10^{16} cm^{-2}; 40 keV) – Pyrrolidine pH=13.12 flow; 45 min

The photolithographic process used to involved a "hard bake" step – heating the wafer to anneal and stabilize resist structures. Applying this hard bake to the implanted resist broke the crust breaking and allowed the solutions to penetrate the resist. So the sample was heated first to 130 °C for 10 s on a hot plate before being treated with a mixture of DI/O_3 + NH_4OH (pH = 13) and UV light. The resist was completely removed and the structured wafer surface damage-free within one hour making this method suitable for high-energy, high-dose implanted PHS-type DUV resists.

Should pyrrolidine finds its way into the semiconductor industry it could be useful to replace NH_4OH with it.

Results and Discussions – Resist stripping

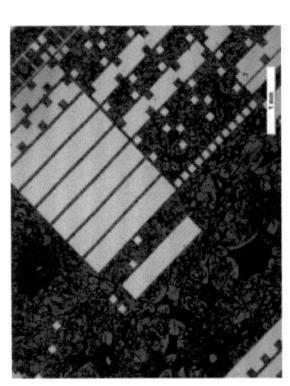

Figure 104 - (As 10^{16} cm^{-2}; 40 keV) - hard bake 130 °C for 10 s

Figure 105 - (As 10^{16} cm^{-2}; 40 keV) – hard bake, DI/O$_3$ + NH$_4$OH (pH=13) flow + UV

A complete overview of the results for pH = 12 - 13.5 on the As 40 keV 10^{16} [cm^{-2}] as images at an optical magnification of 20 is given in the appendix 9.5 in Table XIX.

5. Summary / Zusammenfassung

The main results of the work presented in the foregoing chapters as a PhD thesis are summarized below.

In the production of integrated circuits (ICs), photolithography plays a key role in wafer structuring. The basic principle of photolithography is the selective processing of areas (etching, implantation, metallisation etc.) while the others are covered and therefore protected by the resist. After each process step the resist, now modified, has to be removed. In the history of semiconductor manufacturing this has been accomplished with a mixture of H_2SO_4 and H_2O_2, H_2SO_4 and O_3 or a plasma etch. As the structure sizes decreased they reached a stage where they had to be exposed to light of shorter wavelengths for the photolithography, going from i-line (365 nm) to DUV (248 nm and 193 nm). This change in wavelength now requires new resists and therewith new stripping methods. Beside the changes in the resist the finer structures are also more sensitive to damages caused by the resist strip. Along with this the demand for cost reduction and environment-friendliness poses a big challenge for modern resist stripping. In this study ozone in deionised water (DI/O_3) was the basic chemistry investigated as it is cost efficient in production and disposal as well as environment friendly. Furthermore it is a chemistry known to cause no damage to the wafers. DI/O_3 has been successfully applied to strip i-line resists. The challenge now is to find ways and means to make DI/O_3 strip even highly implanted DUV resists which currently can only be removed by a plasma etch.

To achieve this a detailed understanding of the behaviour of ozone in DI water and the influence of factors both chemical and physical on the stripping efficiency at the different stages in the process is necessary. Along with this, methods which enable the elucidation of resist structures and the changes they undergo during the process of photolithography as well as during the ozone strip have to be developed. This will enable us to understand the mechanisms involved and hence, ideally, develop ozone-based stripping solutions customized for each resist and process step.

For this purpose the ozone decomposition in DI water with and without additives was studied via UV-Vis-spectroscopy. Radicals generated within the ozone decomposition were trapped and quantified, the resists were studied directly on the wafer with IR and Raman spectroscopy and stripped with DI/O_3-mixtures and different setups to find optimum conditions for a complete and damage free resist strip.

Summary

UV-Vis spectroscopy at 260 nm was used to study ozone decomposition and the factors, both chemical and physical, which influence it. These factors are pH, different additives at the same pH, temperature and mixing of the solution. All have been compared to pure DI/O_3 as the reference. From the decomposition curves recorded, the half lifetimes $\tau_{1/2}$ were determined and with them the formal reaction orders n calculated. Based on the reaction orders thus derived the corresponding reaction rates k were calculated. Recording spectra at different temperatures, namely 25 °C and 50 °C, also allowed the calculation of the activation energy E_A for each additive as well as for the different mixing setups.

pH at RT: the half lifetime varies between 5 s for pH = 9 and 110 s for pH = 1 with 102 s for pure DI/O_3 at pH = 3.5 giving a 22-fold increase in the speed of decomposition at pH=9.

Temperature: the maximum influence with an increase by a factor of 1.8 occurs for pure DI/O_3 at pH = 3.5, the minimum influence with the fastest decomposition at pH = 9 with a factor of almost 1. For pH = 1 with H_2SO_4 the increase was only by a factor of 1.2.

Recirculation: in pure DI/O_3: the physical factor recirculation brought about an increase in the speed of decomposition by a factor of 34, in DI/O_3 with H_2SO_4 at pH =1 this factor was 43 and at pH=9 with NH_4OH it was 3.

Therefore, the strongest influence was produced by keeping the solution circulating followed by altering the pH. Increasing the temperature produced the least effect. The reaction orders n for recirculation were close to first order and varied little with the different additives or with temperature. In the stopped flow or quartz cell experiments the reaction order shifted towards second order.

For the radical determination trapping reactions with MeOH and DMSO alone produced good results concerning the linearity over the whole region of interest. They were calibrated by standard addition of predefined amounts of H_2O_2 which on exposure to UV radiation in the region of 310 nm generate OH radicals. The slopes of the calibration curves in both cases increased with the pH of the sample solution resulting in increased radical concentrations being calculated from these curves. Although in both trapping variations radical concentrations increased with increased sample pH their absolute values differed. Iodometry, a self-contained method, was therefore used to verify the results of the MeOH and DMSO trapping method. Iodometry supported the MeOH results although MeOH is claimed to be subject to AOPs.

Summary

IR spectroscopy proved to be a suitable method for the structural characterisation of the resists and the tracking of the changes undergone during the various processing steps as well as the ozone based stripping. Direct measurements in transmission mode through the wafer substrate and the resist were found to be sensitive enough for this purpose. With the basic knowledge of the resist composition it was possible to assign peaks from the spectra to functional groups and then follow the changes by the calculation of differential spectra on samples measured before and after each process step. With this approach it was possible, for example, to follow the removal of the t-Boc protection group during the PEB. The structural changes that take place during the exposure could not be tracked as there were no significant changes in the functional groups. For the stripping with DI/O_3 IR spectroscopy delivered well-defined spectra allowing a good calculation of the differential spectra. These displayed significant peak changes which support the assumption of classical ozonolysis as the decomposition mechanism for the resist. For the study of the resist crust originating from ion implantation IR was fundamentally unsuitable and was replaced by Raman spectroscopy and microscopy.

Raman spectra showed the crust to be of a highly carbon containing structure. Regrettably, the peak assignable to the crust was too broad for the exact composition of the crust to be determined.
The wavelength region of the peak corresponds to that of peaks of glassy carbon and highly ordered and conventional graphite. Such a broad peak suggests that the structure of the crust is not uniform but contains more than one carbon modification.
Although Raman microscopy enables depth profiling by progressively varying the focus, in this case it was not possible to probe the top crust (~ 40 nm) without also probing the unmodified resist below and even the Si-substrate. The least that was possible was a background measurement on a resist free site on the wafer to spectrally compensate for the Si substrate.
As the purpose of all these studies is to enable or improve DI/O_3 based resist stripping on unimplanted as well as high-dose implanted resists the removal efficiency of DI/O_3 spiked with different additives that alter the pH was studied at RT. The experiments with the unimplanted PHS-type DUV resists (JSR M91Y, Rohm&Haas UV26) were conducted in a beaker and the pH varied from 1 to 8. For these unimplanted resists the maximum efficiency could be achieved at pH = 5 – 6. Lowering or increasing the pH beyond this range gave poor results. In case of high pH values the ozone decomposition was too fast for a beaker experiment as the solution needed to be changed within seconds. Resist stripping in M91Y and UV26 was very slow during the steps prior to the PEB (Chapter 4 see 4.6) whereby strip removal in UV26 was slower than in M91Y.

Summary

The reason for this difference lies in the type of protection group (t-Boc in UV26 and unknown in M91Y) as well as in the level of protection. As both resists are of the PHS type, they behave similarly after the PEB during which the protection group is removed. There is then a strong increase in the stripping rate whether or not an additive is used.

Since at higher pH values better results have been obtained for the stripping of unimplated resists, for the implanted DUV248 the setup was changed from a beaker to a dispenser setup allowing an in situ introduction of DI/O$_3$ from outlet that could mix with the additive from a second pipe outlet directly on the wafer. This setup enabled the rapidly decomposing ozone to be continuously renewed with a fresh supply of DI/O$_3$ and the direct application of UV irradiation to advance radical generation. This application of UV radiation produced a significant improvement on the DUV248 implanted with As at a dose of $10^{15} \left[\frac{1}{cm^2} \right]$ and 5 keV enabling it to be completely stripped in 10 min at RT with pure DI/O$_3$. For the implanted resist with the higher dosage of $10^{16} \left[\frac{1}{cm^2} \right]$ at 40 keV the best results were achieved by heating at 130 °C on a hot plate prior to stripping with DI/O$_3$ + NH$_4$OH at pH = 13 with UV irradiation. This allowed a complete strip within an hour. A further, if only slight, increase in pH to 13.5 with KOH resulted in severe damage caused to the silicon substrate by alkaline etching. High-dose implanted resists therefore require long stripping times within a very narrow pH range.

Zusammenfassung

Im Prozeß der Herstellung integrierter Schaltkreise spielt die Photolithographie eine entscheidende Rolle bei der Strukturierung der Wafer. Das Prinzip der Photolithographie beruht dabei auf der selektiven Prozesssierung einiger Bereiche des Wafers (Ätzen, Ionenimplantation, Metallisierung usw.) während andere Bereiche durch den Photolack geschützt werden. Dieser, nun durch die Prozesssierung modifizierte, Photolack muß im Anschluß wieder entfernt werden. In der Geschichte der Halbleiterfertigung geschah dies mit H_2SO_4/H_2O_2, H_2SO_4/O_3, DI/O_3 oder durch Plasmaveraschung. Seitdem die Strukturgröße immer mehr abnimmt, werden immer kürzere Belichtungswellenlängen benötigt, die von i-line (365 nm) bis DUV (248 nm, 193 nm) reichen. Einhergehend mit diesen kürzeren Wellenlängen ist eine notwendige Veränderung der Resist-Struktur und damit die Notwendigkeit neuer Techniken zur Resist-Entfernung. Hinzu kommt, daß diese, nun kleineren Strukturen, ihrerseits empfindlicher gegenüber Schäden aus dem Entfernungsprozeß sind. Als zusätzliche Herausforderung ist der dauerhafte Druck zur Kostenreduzierung sowie zur Umweltverträglichkeit anzusehen. Diese Dissertation beschäftigt sich daher mit der Resist-Entfernung basierend auf ozoniertem Wasser (DI/O_3), da es kostengünstig zu erzeugen und entsorgen ist, gleichzeitig umweltfreundlich und zu guter Letzt dafür bekannt die erzeugten Strukturen nicht zu beschädigen.

Da in der Vergangenheit DI/O_3 schon erfolgreich bei i-line Lacken eingesetzt wurde, besteht nun die Herausforderung darin diese Chemie auch auf moderne und dabei vor allem hochimplantierte DUV-Lacke anzuwenden, die bisher nur mittels Plasmaveraschung entfernbar sind.

Um dieses Ziel erreichen zu können, ist ein detailliertes Verständnis der Ozonchemie in Wasser sowie ihrer Beeinflussung durch Additive wichtig, sowohl in Bezug auf das Ozon selbst als auch den Einfluß auf dessen Lack Entfernungspotential bei verschiedenen Prozeßschritten des Lacks. Einhergehend damit bedarf es Methoden zur Untersuchung der Lackstrukturen und zur Verfolgung ihrer Änderung während bestimmter Prozeßschritte sowie während der Entfernung mittels DI/O_3. Das Ziel ist es das Verhalten von Ozon und Lack zu verstehen, um auf diese Weise Möglichkeiten zur maßgeschneiderten Lackentfernung zu erhalten.

Zu diesem Zweck wurde der Ozonzerfall mittels UV-Vis-Spektroskopie unter verschiedenen Bedingungen (Zusatz, pH, Temperatur, Durchmischung) untersucht. Ferner wurden die, bei der Zersetzung entstehenden, Radikale chemisch abgefangen und quantifiziert. Der Photolack selbst wurde direkt auf dem Wafer mittels IR- und Raman-Spektroskopie untersucht und zu guter Letzt dessen Entfernung mit DI/O_3 und verschiedensten Zusätzen sowie Versuchsaufbauten in Angriff genommen.

Zusammenfassung

Dieses Kapitel soll nun einen zusammenfassenden Überblick über die wichtigsten Ergebnisse dieser Dissertation geben.

Bei der Beobachtung und Quantifizierung der Ozonzersetzung mittels UV-Vis-Spektroskopie bei 260 nm wurden verschiedene Faktoren, sowohl chemischer als auch physikalischer Natur, auf ihren Einfluß auf die Ozonzersetzung in Wasser hin untersucht. Dabei handelt es sich um den pH-Wert, die Natur verschiedener Additive bei gleichem pH-Wert, die Temperatur, sowie den Einfluß von Mischungseffekten bei der Verwendung von Additiven. Als Referenz diente jeweils pures DI/O_3.

Aus den gewonnenen Zerfallskurven wurden zuerst die Halbwertszeiten $\tau_{1/2}$ ermittelt und aus deren Verhältnis die formale Reaktionsordnung n, sowie die dazu gehörige Reaktionskonstante k bestimmt. Ferner wurde durch Messungen bei 25 °C und 50 °C die zu jedem Additiv gehörige Aktivierungsenergie E_A errechnet.

Betrachtet man die pH-Abhängigkeit der Zersetzung bei RT so variieren die Halbwertszeiten zwischen 5 s bei pH = 9, 110 s bei pH = 1 und 102 s für pures DI/O_3 bei pH = 3,5. Dies macht einen maximalen Zeitfaktor von 22 aus.

Betrachtet man dagegen den Einfluß der Temperatur auf den Zerfall so macht eine Temperaturerhöhung von 25 °C auf 50 °C nur einen maximalen Zeitfaktor von 1,8 aus, diesmal bei purem DI/O_3 und von nur 1.2 bei pH = 1 mit dem Zusatz von H_2SO_4. Für hohe pH-Werte pH = 9 ist der Temperatureinfluß nicht mehr feststellbar.

Überraschender Weise stellt sich der physikalische Effekt des Durchmischens als der stärkste Einfluß heraus, da er einen maximalen Zeitfaktor von 43 für pH = 1 mit H_2SO_4 liefert. Selbst ohne das Vorhandensein eines Additives in purem DI/O_3 beträgt der Zeitfaktor noch 34 und wird erst bei hohem pH-Wert mit einem Wert von 3 bei pH = 9 merklich kleiner.

Somit stellt das Durchmischen von allen Einflüssen auf die Ozonzersetzung den stärksten dar.

Die bestimmten Reaktionsordnungen sind größtenteils unabhängig von den Parametern und bewegen sich nahe erster Ordnung. Einzig allein die Variante „stopped flow" führt zu einer Verschiebung Richtung zweite Ordnung.

Zusammenfassung

Im Abschnitt der Radikalquantifizierung zeigen sowohl die Abfangreaktion mit MeOH als auch mit DMSO für sich allein gute Resultate. Beide weisen eine gute Linearität der Kalibrierkurven über den gesamten untersuchten Bereich auf und lassen sich dabei exakt kalibrieren. Die Kalibrierung gelingt bei beiden Methoden mittels Standardaddition von OH-Radikalen, welche aus H_2O_2 mittels UV-Bestrahlung bei 310 nm erzeugt werden. Bei beiden Abfangmethoden läßt sich eine Erhöhung der Steigung der Kalibriergeraden mit dem pH-Wert der Probenlösung beobachten. Die aus diesen Kurven errechneten Radikalkonzentrationen der Proben steigen dabei ebenfalls mit dem pH-Wert an. Obwohl beide Methoden dieselbe Tendenz aufweisen, unterscheiden sich ihre absoluten Werte allerdings deutlich. Um dies zu verifizieren wurde auf die klassische Iodometrie als eine Absolutmethode zurück gegriffen. Diese Methode neigt dazu die Ergebnisse der Abfangreaktion mittels MeOH zu bestätigen, obwohl diese dafür bekannt ist für „advanced oxidations processes" (AOPs) anfällig zu sein, durch die die Ergebnisse zu höheren Radikalkonzentrationen verfälscht werden.

Für das Ziel der strukturellen Analyse des Photolackes sowie der Verfolgung von Änderungen während verschiedener Prozessierungsschritte sowie der Ozonbehandlung, hat sich die IR-Spektroskopie als geeignetes Mittel erwiesen. Entgegen der Befürchtungen hat sich die einfache Messung in Transmission, direkt durch den Wafer und den Lack hindurch, als ausreichend empfindlich erwiesen. So war es möglich, ausgehend von dem Wissen über die Monomerbestandteile des Lackes, die gefundenen Peaks zu funktionellen Gruppen zu zuordnen. Mittels Differenzspektren konnte z.B. die Abspaltung der Schutzgruppe (t-Boc) während des „post exposure bakes" (PEB) beobachtet werden. Besonders hilfreich war die IR-Spektroskopie bei der Analyse der Behandlung des Lacks mit DI/O_3, wobei klare Hinweise auf die Ozonolyse als Zersetzungsmechanismus gefunden wurden. Leider gilt es anzumerken, daß die IR-Spektroskopie keinerlei brauchbare Informationen zur Art, Umwandlung und Freisetzung der Photosäure liefert und bei der Anwendung auf implantierten Lacken gänzlich versagt.

Die Anwendung von Raman-Spektroskopie, als komplementärer Methode zur IR-Spektroskopie, für die Analyse der Kruste bei implantierten Photolacken ermöglicht eine Erkennung selbiger, von der angenommen wird, daß es sich um hochvernetzten Kohlenstoff handelt. Leider ist der dazugehörige Raman-Peak derart breit, daß es nicht möglich ist ihn eindeutig zu zuordnen. Er legt allerdings eine Mischung aus verschiedenen Graphitmodifikationen nah. Der Versuch mittels Raman-Mikroskopie durch eine entsprechende Fokussierung ausschließlich die Kruste (obersten ~ 40 nm) zu messen gelang leider nicht, da unter jedweder Fokussierung auch immer das Si-Substrat detektiert wurde. Das Si-Substrat stellt allerdings kein Problem dar, da es sich im Spektrum, mittels Differenzmessung, leicht heraus rechnen läßt.

Zusammenfassung

Die Untersuchung des Entfernungsprozesses mittels DI/O_3 und die Evaluierung verschiedener Additive stellt den für die praktische Anwendung wichtigsten Teil dar. Als nicht implantierte Proben wurden dazu die Lacke M91Y von JSR Micro sowie UV26 von Rohm&Haas verwendet und die Versuche im Becherglas durchgeführt. Bei der Variation des pH-Werts im Bereich pH = 1 bis pH = 9 mittels verschiedener Additive, konnte für beide Lacke in allen Prozeßschritten ein Optimum im Bereich 5 – 6 mit HAc- oder Phosphatpuffer ermittelt werden. Ferner konnte bei beiden Lacken eine sehr große Abhängigkeit vom Vorhandensein der Schutzgruppe bzw. deren Fehlen nach dem PEB festgestellt werden, wobei die Entfernung eine deutliche Erhöhung der Ablösegeschwindigkeit bewirkt. Vor dem PEB, d.h. solange die Schutzgruppen noch nicht entfernt wurden, lies sich der M91Y besser entfernen als der UV26. Danach existierte kein Unterschied da es sich bei beiden um Polyhydroxystyrol (PHS) basierte Lacke handelt, die sich nur in der Art der Schutzgruppe und dem Grad der Schützung unterscheiden.

Da sich für nicht implantierte Lacke hohe pH-Werte als besser erwiesen haben, einhergehend damit allerdings ein schneller Ozonzerfall auftritt, wurde für die Versuche mit implantierten Lacken der Aufbau in Richtung eines „single-wafer tools" mit Aufsprühdüsen variiert, so daß das Mischen von DI/O_3 und Additiven kontinuierlich auf dem Wafer erfolgen kann. Mit diesem Aufbau war es möglich den Lack mit der niedrigeren Arsen-Dosis von $10^{15} \left[\frac{1}{cm^2} \right]$ und einer Implantationsenergie von 5 keV durch eine Kombination von DI/O_3 und UV-Bestrahlung des Wafers bei RT innerhalb von 10 min rückstandslos und ohne Schäden auf dem Wafer zu entfernen. Selbiger Ansatz für die höhere Dosis von $10^{16} \left[\frac{1}{cm^2} \right]$ bei 40 keV zeigte keinerlei erkennbaren Erfolg. Erst die Anwendung von sehr hohen pH-Werten, erzeugt mittels NH_4OH, KOH oder Pyrrolidin, konnte Erfolge aufweisen. Das Prozeßfenster dafür war allerdings sehr gering, da pH-Werte von 12 oder kleiner keinerlei Entfernung bewirkten, zu hohe pH-Werte von über 13 bei der nötigen Prozeßzeit von einer Stunde allerdings die Waferoberfläche beschädigten. Die besten Ergebnisse konnten bei einem pH-Wert von 13 mit der Kombination von DI/O_3 + NH_4OH + UV erzielt werden. Damit war es möglich den $10^{16} \left[\frac{1}{cm^2} \right]$ implantierten Lack innerhalb von einer Stunde ohne Beschädigung der Waferoberfläche vollständig zu entfernen.

6. Outlook

On the basis of the knowledge and insight gained from this study a proposal for the course of further work in this direction is given below.

The method for the detection (by UV spectroscopy), recording and monitoring of ozone decomposition does not, in my point of view, require further development. The main factors that influence ozone decomposition have been determined in this study. However, when new and better additives are to be used to increase stripping efficiency, it could be useful to determine exactly the rate of ozone decomposition and other relevant parameters rather than to deduce them from the trends pointed to in this study.

The questions that arose on determining the concentration of radicals during ozone decomposition could not all be answered satisfactorily within the scope of this work. Why, for instance, do the results obtained with MeOH trapping differ from those obtained with DMSO? It would be good to find out why. Another approach would be to develop the iodometric titration procedure to cover the full range of pH values in water, or to find an alternative self-contained method that could serve to clarify the different values obtained with MeOH and the DMSO method.

IR spectra in transmission mode have proven to be useful for the structural characterization of unimplanted resists. A change in the mode of measurement to direct reflection from the sample or with ATR could increase the sensitivity of the method. Another approach would be to devise a setup to follow in real time the changes that occur in the stripping process.

For Raman microscopy the resist crust was too thin for a progressive change in focus which would have allowed a depth profile and a better picture of the various modifications of the carbon of which the crust was made up. Improved spectral peak fitting and interpretation would be useful, so would an improvement in the measurement setup to improve resolution.

Outlook

The last aspect, the one for which everything else has been done is the resist stripping. For the unimplanted PHS-type resists studied the most imported task is to test the best results from the beaker setup with a dispenser setup for a hopefully positive effect on the stripping efficiency. For the implanted PHS-type resist organic bases should also be tried out all in combination with a thermal pretreatment and with UV irradiation with the dispenser setup. This could shorten the stripping time thus preventing damage to the substrate, and allow the pH to be reduced reduce to an acceptable level. Another effect worth following up is the in situ heating of the sample during the strip as a way to generate more radicals and in general to increase the kinetics of the process. This might be achieved for instance with an additional IR lamp to the dispenser setup. For all stripping studies it is essential to transfer the best results from the laboratory setup to an industrial scale up to test their applicability. Furthermore this transfer would provide the scope for new parameters like the DI/O_3 as well as the flow of additives or the rotation speed of the wafer.

7. Literature references

1. **Zell, T.** Vortrag: "Lithography". 7. Dresdner Sommerschule Mikroelectronik : s.n., 2006.

2. **R.W. Cahn, P. Haasen, E.J. Kramer.** *Material Science and Technology - Processing of Semiconductors.* s.l. : VCH Verlagsgesellschaft mbH, D-69451 Weinheim, 1996. S. 198-207. Bd. 16. 3-527-26829-4.

3. **Cahn, P. Haasen, E.J. Kramer.** *Material Science and Technology - Processing of Semiconductors.* s.l. : VCH Verlagsgesellschaft mbH, D-69451 Weinheim, 1996. S. 207-227. Bd. 16. 3-527-26813-8.

4. **Myers, Andrew G.** Homepage Prof. Andrew G. Myers. [Online] Harvard University. [Zitat vom: 08. 04 2009.] http://www.chem.harvard.edu/groups/myers/handouts/3_Protective.pdf.

5. **Clayden, Greeves, Warren, Worthers.** *Organic Chemistry.* s.l. : Oxford University Press, 2001. S. 655.

6. **Douki, Katsuji.** High-Performance 193-nm Positive Resist Using Alternating Polymer System of Functionalized Cayclic Olefins / Maleic Anhydride. [Hrsg.] Fine Electronic Reasearch Laboratories JSR Corporation Semiconductor Materials Laboratory.

7. **Leitzke, Achim.** Dissertation. [Hrsg.] Universität Duisburg-Essen. *Mechanistische und kinetische Untersuchungen zur Ozonolyse von organischen Verbindungen in wä ssriger Lösung.* 2003.

8. *Presentation - Ozonized water is green.* **Knotter, Martin.** UCPSS 2008 Brügge - Tutorials : s.n., 2008.

9. **Vankerckhoven, Hans.** Dissertation. *Fundamental Study of the Degradation of Organic Compounds by Ozone/Water Processes: Application in Semiconductor Cleaning.* Katholieke Universiteit Leuven : s.n., 2004.

10. Henry's law constants. *Max-Planck-Institut für Chemie.* [Online] http://www.mpch-mainz.mpg.de/~sander/res/henry.htm.

11. **Weast, Robert C.** *CRC Handbook of Chemistry and Physics 63rd Edition.* s.l. : CRC Press, Inc. Boca Rato, Florida, 1982-1983. 0-8493-0463-6.

12. **Sauerstoff.** Wikipedia. [Online] http://de.wikipedia.org/wiki/Sauerstoff.

13. **Hart, Edwln J.** Molar Absorptivities of Ultraviolet and Visible Bands of Ozone in Aqueous Solutions. *Anal. Chem.* 1983, Bd. 55, S. 46-49.

14. **Clayden, Greeves, Warren, Worthers.** *Organic Chemistry.* s.l. : Oxford University Press, 2001. S. 938-939.

Literature references

15. **Sehested, Knud.** The Primary Reaction in the Decomposition of Ozone in Acidic Aqueous Solutions. *Environ. Sci. Technol.* 1991, 25, 1589-1596.

16. **Virdis, A.** A novel kinetic mechanism of aqueous-phase ozone decomposition. *Ann. Chim. (Roma).* 1995, 85, 633-647.

17. **Tomiyasu.** Kinetics and Machanism of Ozone Decomposition in Basic Aqueous Solutions. *Inorg. Chem.* 1985, 24, 2962-2966.

18. **Clayden, Greeves, Warren, Worthers.** *Organic Chemistry.* s.l. : Oxford University Press, 2001. S. 1021-1051.

19. **Nash, T.** The Colorimetric Estimation of Formaldehyde by Means of the Hantzsch Reaction. *Biochem J.* . 1953, 55(3), 416-421.

20. **Lackhoff, Marion.** Dissertation. [Hrsg.] Technischen Universität München. *Photokatalytische Aktivität ambienter Partikelsysteme.* Technischen Universität München : s.n., 2002. S. 11.

21. **Tai, Chao.** Determination of hydroxyl radicals in advanced oxidation processes with dimethyl sulfoxide trapping and liquid chromatography. *Analytica Chemica Acta.* 2004, 527, 73-80.

22. **Ulanski, Piotr.** The OH radical-induced chain reactions of methanol with hydrogen peroxide and with peroxodisulfate. *J. Chem. Soc., Perkin Trans. 2.* 1999, 165-168.

23. **Perkin Elmer.** [Online] [Zitat vom: 05. Mai 2009.] http://las.perkinelmer.com/content/TechnicalInfo/TCH_FTIRATR.pdf.

24. **Hesse, Meier, Zeeh.** *Spektroskopische Methoden in der organischen Chemie.*

25. **Römpp Online.** [Online] Thieme. [Zitat vom:] http://www.roempp.com/prod/index1.html.

26. **Hoffmann, Günter G.** Micro-Raman and Tip-Enhanced Raman Spectroscopy of Carbon Allotropes. *Macromol. Symp.* 2008, 265, 1-11.

27. **Gottschalk, Christiane.** *Oxidation organischer Mikroverunreinigungen in natürlichen und synthetischen Wässern mit Ozon und Ozon/Wasserstoffperoxid.* Berlin : Shaker Verlag Aachen, 1997. S. 51-52. 3-8265-2430-6.

28. **Upadhyay, Santosh K.** *Chemical Kinetics and Reaction and Dynamics.* s.l. : Springer, 2006. ISBN 1402045468, 9781402045462.

29. **Wedler, Gerd.** *Lehrbuch der Physikalischen Chemie.* s.l. : Wiley-VCH Verlagsgesellschaft, D-69469, 1997. 3-527-29481-3.

30. **National Institute of Standards and Technology.** NIST Chemistry WebBook. [Online] http://webbook.nist.gov/chemistry/.

31. **National Institute of Advanced Industrial Science and Technology.** Spectral Database for Organic Compounds - SDBS. [Online] http://riodb01.ibase.aist.go.jp/sdbs/cgi-bin/cre_index.cgi?lang=eng.

Literature references

32. **University Erlangen/Germany.** TelecSpec - IR Spectra Simulation on the WorldWideWeb. [Online] http://www2.ccc.uni-erlangen.de/services/telespec/.

33. [Online] http://www.science-and-fun.de/tools/.

34. **IMEC.** *Information from Rita Vos.*

35. Wissenschaft-technik-ethik.de. [Online] [Zitat vom: 8. 04 2009.] http://www.wissenschaft-technik-ethik.de/wasser_ph.html.

36. **Sotelo, José L.** Ozone Decomposition in Water: Kinetic Stud. *Ind. Eng. Chem. Res.* 1987, 26, 39 - 43.

37. **Beltrán, Fernando J.** *Ozone reaction kinetics for water and wastewater systems.* s.l. : CRC Press LCC, 2004. 1-56670-629-7.

38. **Gottschalk, Christiane.** *Oxidation organischer Mikroverunreinigungen in natürlichen und synthetischen Wässern mit Ozon und Ozon/Wasserstoffperoxid.* Berlin : Shaker Verlag, 1997. S. 25. 3-82652430-6.

39. **Cséfalvay, Edit.** Modelling of wastewater ozonation – determination of reaction kinetic constants and effect of temperature. *Chemical Engineering.* 2007, 51/2, 13-17.

40. **LJ Takic, et al.** A study on the kinetics of ozone decomposition in water of different quality. *Chem. Ind.* 2004, 58(3), 118-122.

41. **Lenntech Water- & Luchtbeh. Holding b.v.** Lenntech Water- & Luchtbeh. Holding b.v. [Online] [Zitat vom: 07. 09 2009.] http://www.lenntech.com/ozone/ozone-decomposition.htm.

42. **Fábián, István.** Reactive intermediates in aqueous ozone decomposition: A mechanistic approach. *Pure Appl. Chem.* 78, 2006, Bd. 8, S. 1559–1570.

43. **IMEC, Rita Vos at.** IMEC - Research center in nano-electronics and nano-technology. *Kapeldreef 75; B-3001 Leuven; Belgium.*

44. **J.G. Calvert, J.N. Pitts Jr.** *Photochemistry.* s.l. : Wiley, 1966.

45. **Karen Reinhard, Werner Kern.** *Handbook of SILICON WAFER CLEANING TECHNOLOGY.* Secind Edition. Norwich, NY 13815 : William Andrew Inc., 2008. 978-0-8155-1554-8 (978-0-8155).

46. *Römpp - Lexikon der Chemie.*

8. Chemicals and equipment

- acetic acid 100 % (CH_3COOH); Fischer scientific analytical reagent grade; product: A/0400/PB17

- phosphoric acid 85 % (H_3PO_4); Grüssing for analytical purposes; product: 13032

- ammonium hydroxide 28 % (NH_4OH); Merck SLSI Selectipur; product: 1.11868.2500

- formic acid 99 %, pure (CHOOH); Acros Organics; product: 147930010

- propionic acid 99 %, pure (CH_3CH_2COOH); Acros Organics; product: 149300010

- sulphuric acid 96 % (H_2SO_4); Merck VLSI Selectipur; product: 1.00709.2500

- nitric acid 65 % (HNO_3); Merck pro analysi; product: 1.00456.1000

- hydrofluoric acid 50 % HF; BASF VLSI Selectipur; product: 51151083

- deionized water/DIW (semiconductor quality)

- oxygen (O_2); oxygen 2.5 compressed 200 bar; Air Liquide

- ozone (O_3)

- sodium hydroxide 99 % (NaOH); Grüssing 99 % reinst; product: 12156

- potassium hydroxide 85 % (KOH); Grüssing for analytical purposes; product: 12038

- sodium carbonate 99.5 % (Na_2CO_3); Grüssing for analytical purposes; product: 12117

- potassium hydrocarbonate ($KHCO_3$); Merck pro analysi; product: 4845

Chemicals and equipment

- formaldehyde 36 % (CH_2O); VWR Rectapur; product: 20910.294

- methanol 100 % (CH_3OH); Merck KGAA p.a. ACS,ISO,Reag. Ph Eur 1L; Product1.06009.1011

- dimethyl sulfoxide/DMSO 99.9% (C_2H_6SO); Sigma-Aldrich A.C.S. reagent; product: 47230-100 mL

- acetylacetone 99.5 % ($C_5H_8O_2$); Fluka puriss p.a. (GC); product: 00900

- ammonium acetate (NH_4CH_3COO); Merck Fractopur; product: 1.16103.1000

- hydrogen peroxide 30 % (H_2O_2); Ashland Chemical; GIGABIT quality

- potassium iodide briquettes 99 % (KI); Acros Organics; product: 196735000

- sodium thiosulfate 0.1 M ($Na_2S_2O_3$); Roth ROTI VOLUM for volumetric solutions; product: CN55.1

- hex-ammonium-hepta-molybdate $(NH_4)_6Mo_7O_{24} \cdot 4H_2O$; Merck KGAA krist. reinst 250 g; Product: 1.01180.0250

Chemicals and equipment - Equipment

- ozone module from SEZ; Astex SEMOZON 09.2 ozone generator

- PerkinElmer Lambda 25 UV/Vis spectrometer with peristaltic sipper

- UV/Vis-Photometer; Merck SQ118 V1.21

- PerkinElmer Spectroscopy Flow-Through Cell, Semi-Micro/Ultra-Micro with In- and Outlet Tubes, Quartz SUPRASIL®, Light Path: 10 mm

- Hellma liquid cell, Quartz Suprasil®, Light Path: 10 mm, Type-Nr. 114-QS

- Roth Plastibrand®, PS 2.5 mL macro disposable cuvettes

- Roth 15 mL polypropylene centrifuge tubes with screw caps

- Julabo MB-5A heating circulator with open bath

- Jasco FT/IR 470 Plus Spectrometer

- pH-electrode pH90 by WTW

- Hamamatsu Lightningcure™ L8444-02 (LC4); 3500mW/cm^2; λ=240 - 400 nm

- Plasmos SD Series Ellipsometer (633 nm)

9. Appendix

9.1. Detailed theory for negative resists

AHR-resists

In the AHRs the two main parts beside the polymer itself are a) the photo acid generators (PAGs - Figure 106) which is the photo active component (PAC) and is either a salt (DPI – diphenyl iodinium, TPS – triphenyl sulfonium) or a non-ionic substance, and b) the crosslinkers (Figure 107) which are bi-functional molecules responsible for the decreased solubility after irradiation.

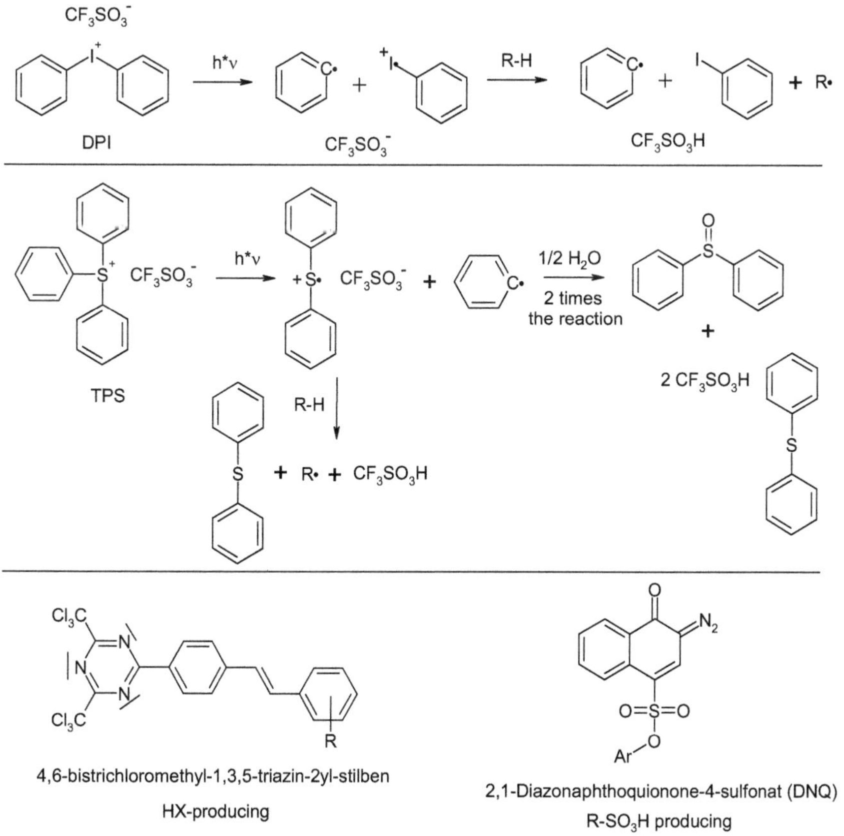
$CF_3SO_3^-$
h*v
R-H
+ R•
DPI
$CF_3SO_3^-$
CF_3SO_3H
$CF_3SO_3^-$
h*v
+S• $CF_3SO_3^-$ +
1/2 H_2O
2 times
the reaction
TPS
R-H
+ R• + CF_3SO_3H
2 CF_3SO_3H
4,6-bistrichloromethyl-1,3,5-triazin-2yl-stilben
HX-producing
2,1-Diazonaphthoquionone-4-sulfonat (DNQ)
R-SO_3H producing

Figure 106 - Photo acid generators (PAG's)

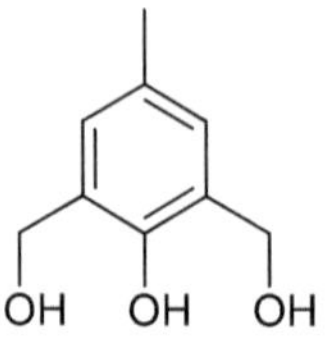

Appendix - Theory

Figure 107 - Crosslinkers for AHR resists

- 2,6-Dihydroxymethyl-4-methylphenol
- Hexamethoxymethyl melamine (HMMM) 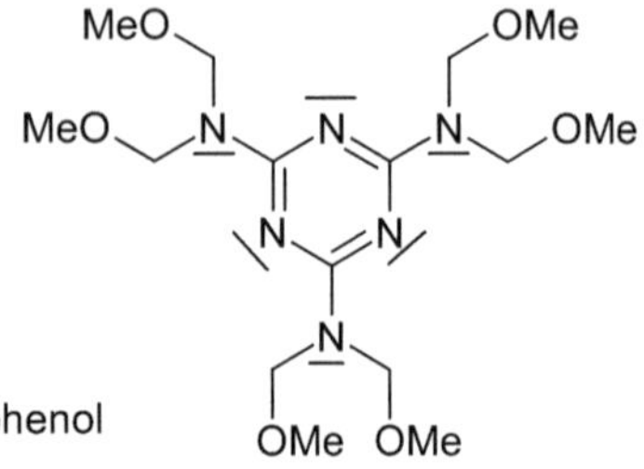
- 4,4'-Dimethoxymethyl diphenyl ether 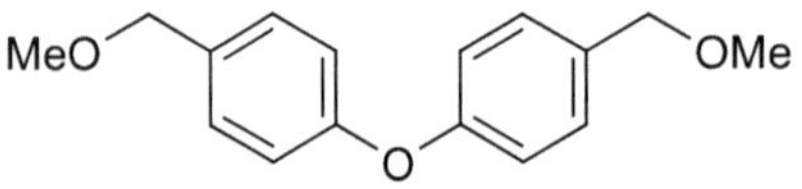
- 1,3,5-Triacetoxymethyl diphenyl ether 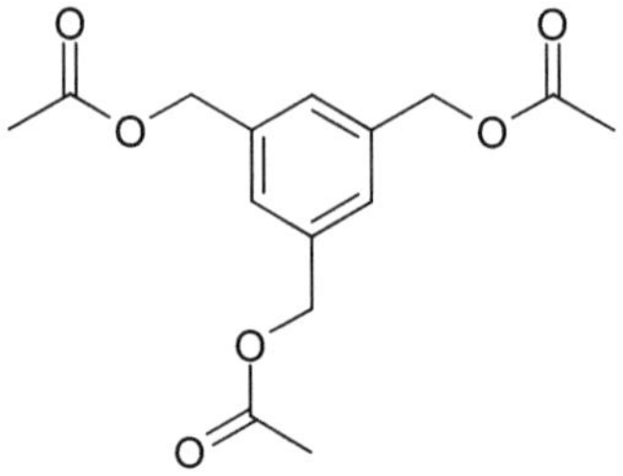

The PAG is evenly dispersed within the resist and designed to absorb the light used for lithography. During the lithography step the PAG precursor is converted to the acid which is then released. This step is followed by the post exposure bake (PEB) which allows the acid to catalytically initiate the crosslinking of the polymers (Figure 108).

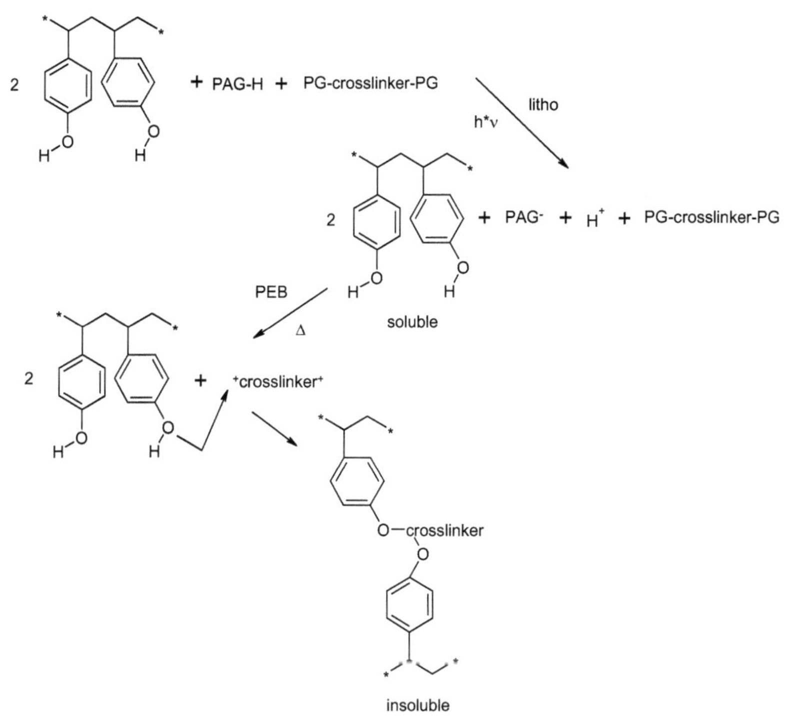

Figure 108 - Mechanism of acid hardening in resists

Appendix - Theory

Radical polymerisation

In practical applications the radical starters can be for example of the Norrish type I (Figure 110) or type II (Figure 111) [44].

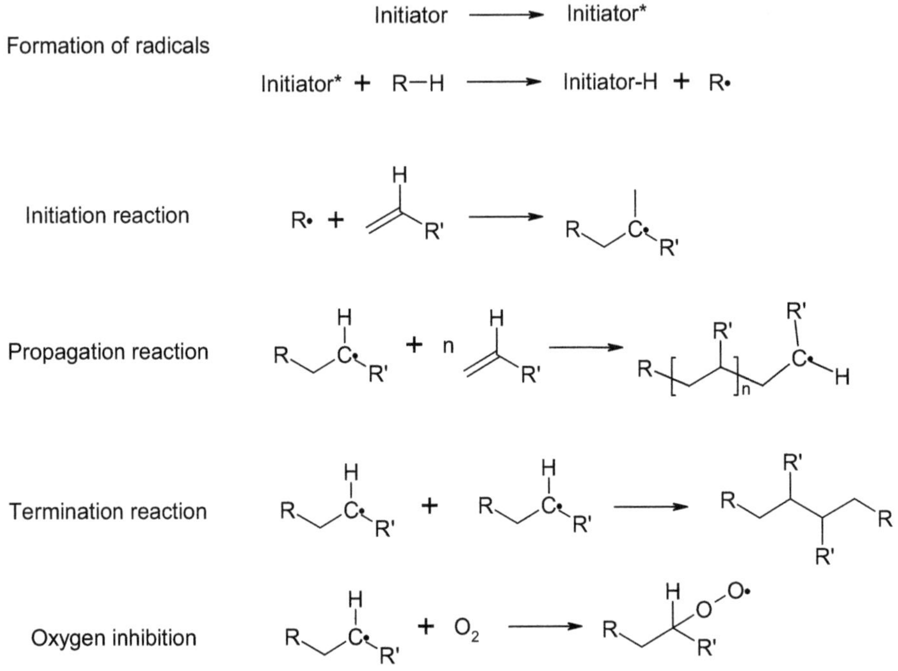

Formation of radicals

$$\text{Initiator} \longrightarrow \text{Initiator*}$$

$$\text{Initiator*} + \text{R}-\text{H} \longrightarrow \text{Initiator-H} + \text{R}\cdot$$

Initiation reaction

$$\text{R}\cdot + \text{CH}_2=\text{CHR'} \longrightarrow \text{R}-\text{CH}_2-\overset{\cdot}{\text{C}}\text{HR'}$$

Propagation reaction

$$\text{R}-\text{CH}_2-\overset{\cdot}{\text{C}}\text{HR'} + n\,\text{CH}_2=\text{CHR'} \longrightarrow \text{R}-[\text{CH}_2-\text{CHR'}]_n-\text{CH}_2-\overset{\cdot}{\text{C}}\text{HR'}$$

Termination reaction

$$\text{R}-\text{CH}_2-\overset{\cdot}{\text{C}}\text{HR'} + \text{R}-\text{CH}_2-\overset{\cdot}{\text{C}}\text{HR'} \longrightarrow \text{R}-\text{CH}_2-\text{CHR'}-\text{CHR'}-\text{CH}_2-\text{R}$$

Oxygen inhibition

$$\text{R}-\text{CH}_2-\overset{\cdot}{\text{C}}\text{HR'} + \text{O}_2 \longrightarrow \text{R}-\text{CH}_2-\text{CHR'}(\text{OO}\cdot)$$

Figure 109 - General mechanism of photopolymerization

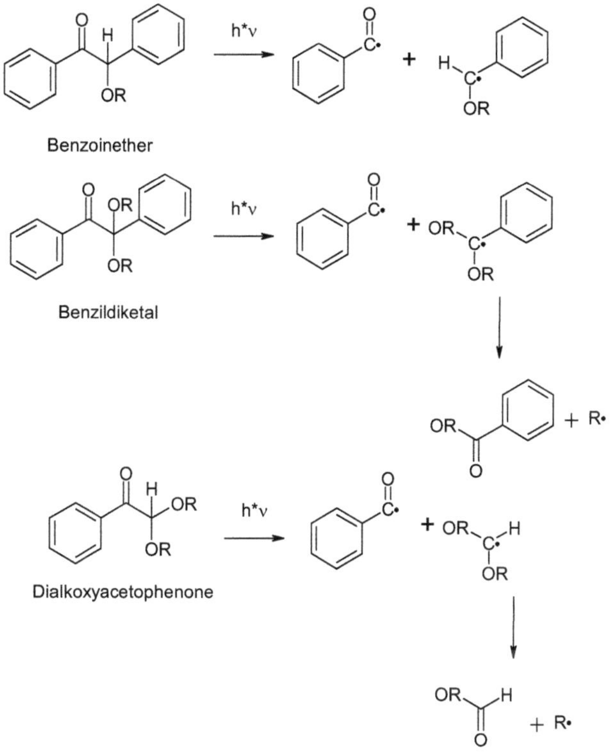

Appendix - Theory

Figure 110 - Norrish type I radical starters

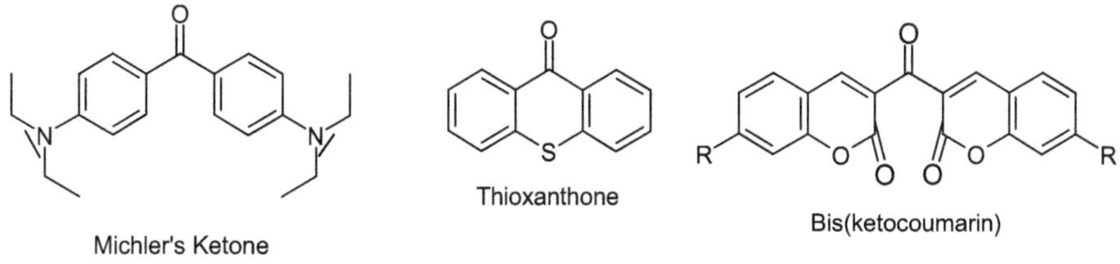

(Benzoinether, Benzildiketal, Dialkoxyacetophenone)

Figure 111 - Norrish type II radical starters

(Michler's Ketone, Thioxanthone, Bis(ketocoumarin))

Anionic and cationic polymerisation

A good example of a cationic polymerization is the Lewis acid induced photopolymerization with BF_3 as a ring opening polymerisation with epoxide monomers as in the SU-8 epoxy resin (Figure 10).

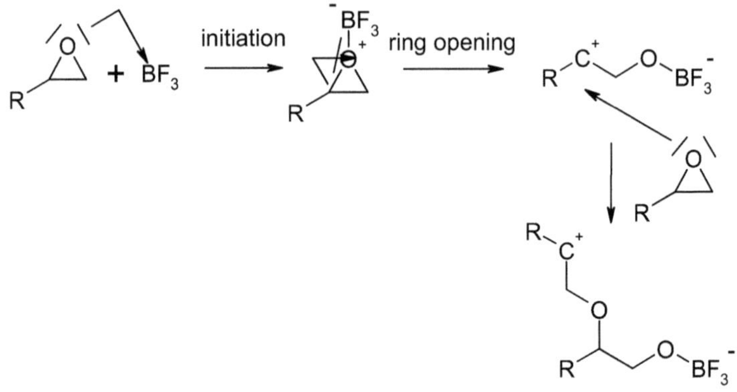

Figure 112 - Lewis acid induced cationic ring opening polymerzation

Appendix - Results

9.2. UV/Vis spectroscopic determination of ozone decomposition

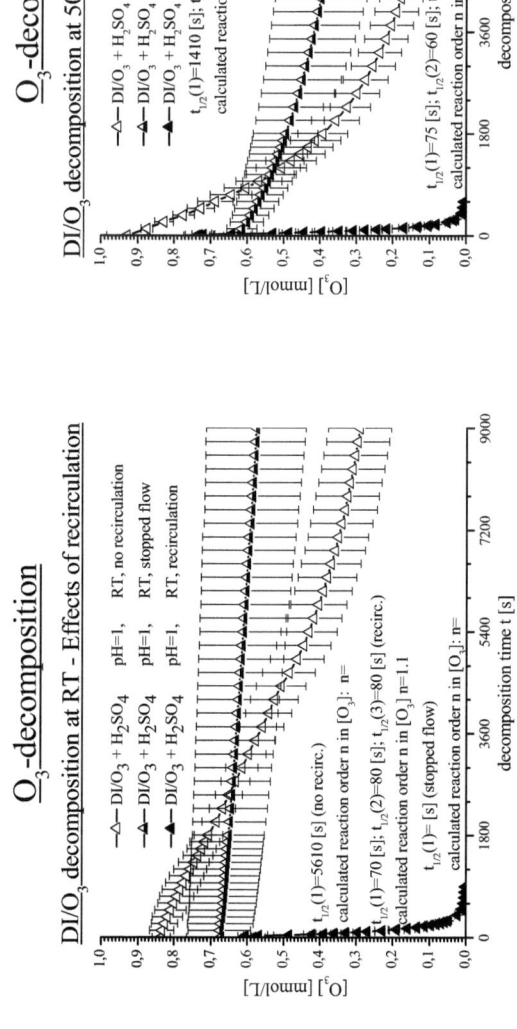

Figure 113 - [O_3] vs. t; RT; mixing effects; pH=1; H_2SO_4

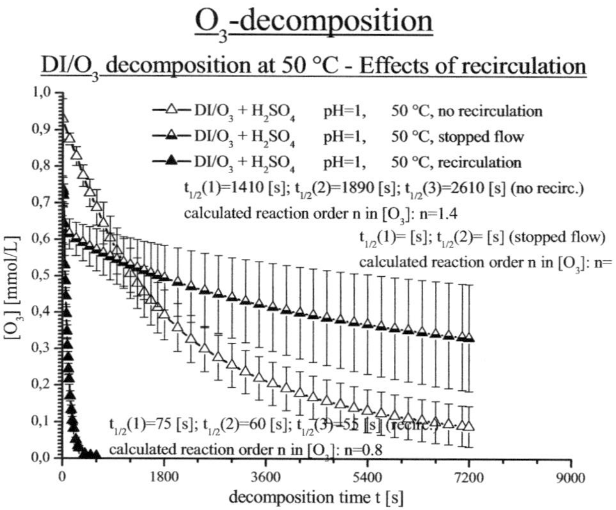

Figure 114 - [O_3] vs. t; 50 °C; mixing effects; pH=1; H_2SO_4

Appendix - Results

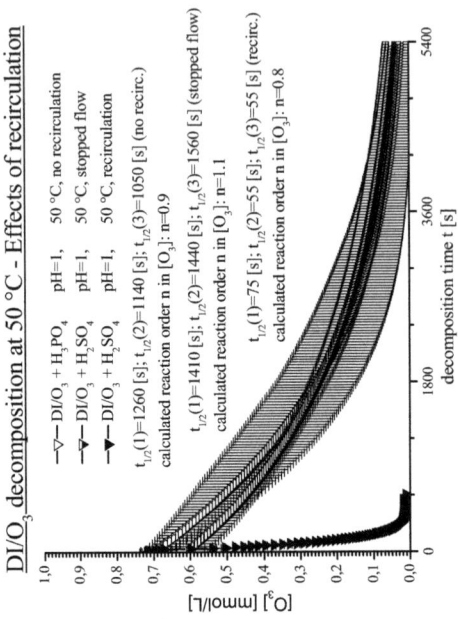

Figure 116 - [O₃] vs. t; 50 °C; mixing effects; pH=1; H₃PO₄

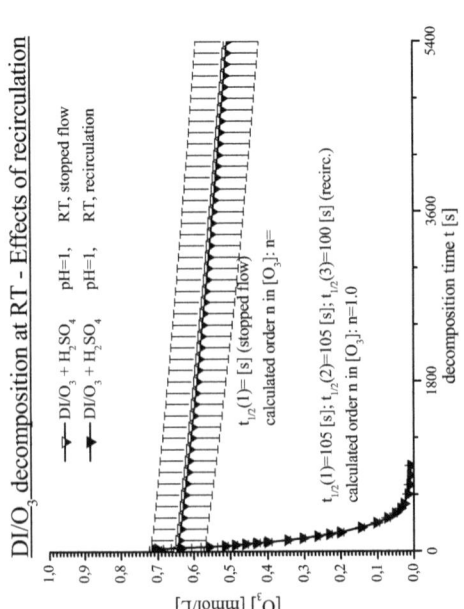

Figure 115 - [O₃] vs. t; RT; mixing effects; pH=1; H₃PO₄

Appendix - Results

O_3-decomposition

DI/O_3 decomposition at RT - Effects of recirculation

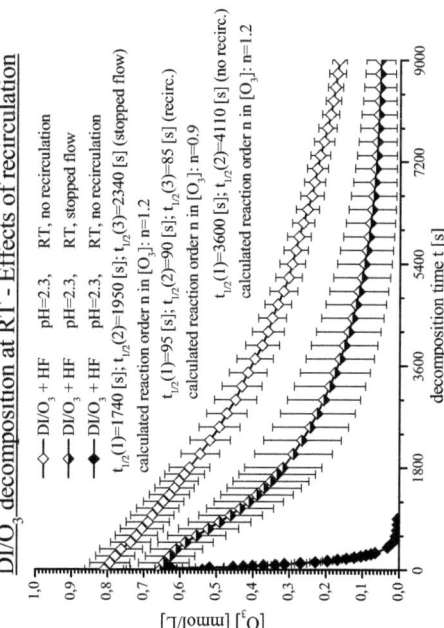

Figure 117 - $[O_3]$ vs. t; RT; mixing effects; pH=2.3; HF

O_3-decomposition

DI/O_3 decomposition at 50 °C - Effects of recirculation

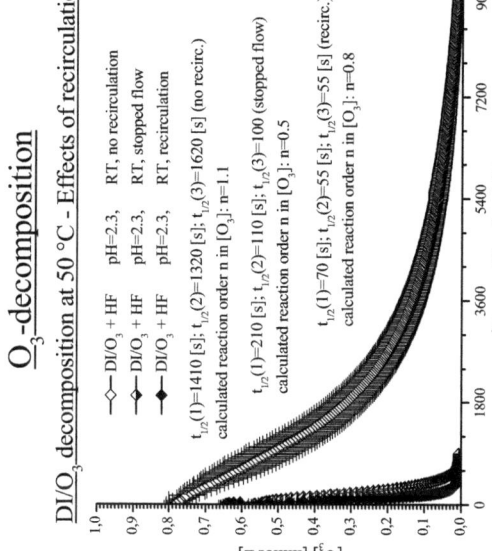

Figure 118 - $[O_3]$ vs. t; 50 °C; mixing effects; pH=2.3; HF

Appendix - Results

O$_3$-decomposition

DI/O$_3$ decomposition at RT - Effects of recirculation

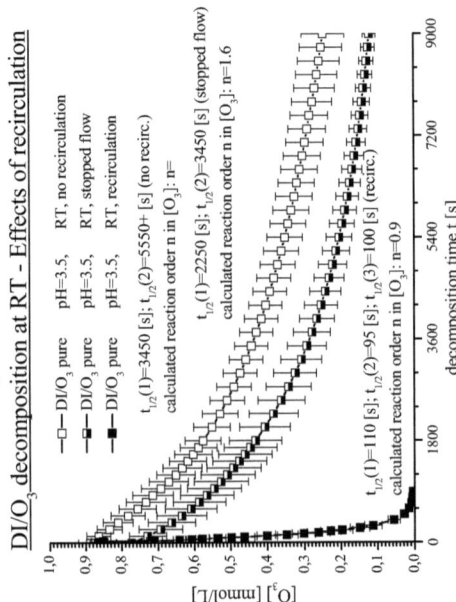

Figure 119 - [O$_3$] vs. t; RT; mixing effects; pH=3.5; pure

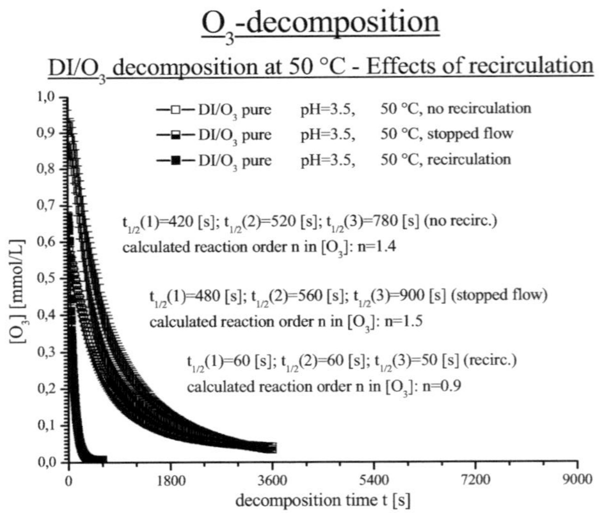

O$_3$-decomposition

DI/O$_3$ decomposition at 50 °C - Effects of recirculation

Figure 120 - [O$_3$] vs. t; 50 °C; mixing effects; pH=1; pure

Appendix - Results

O$_3$-decomposition
DI/O$_3$ decomposition at RT - Effects of recirculation

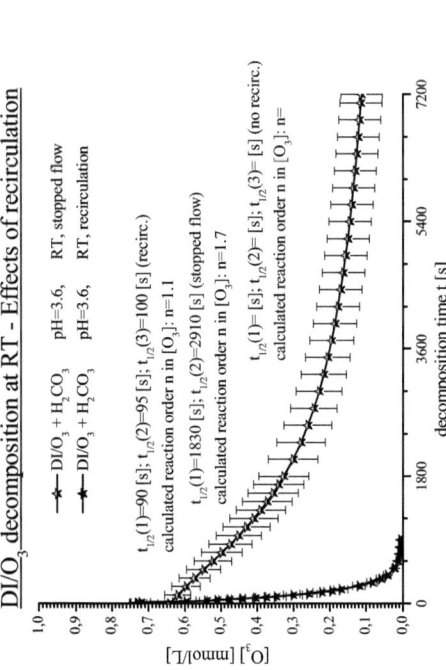

Figure 121 - [O$_3$] vs. t; RT; mixing effects; pH=3.6; H$_2$CO$_3$

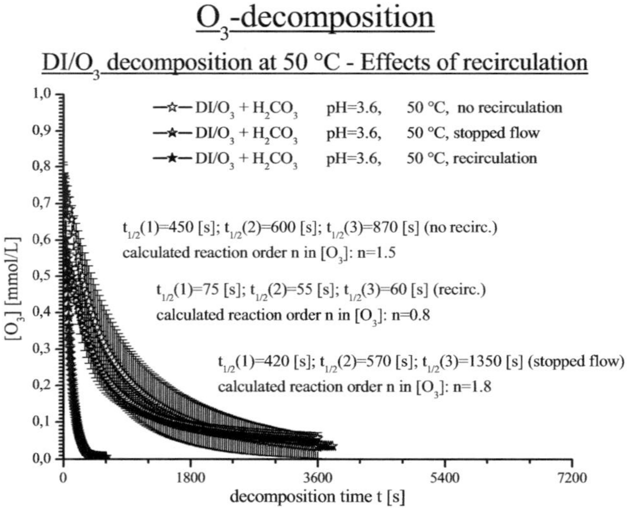

Figure 122 - [O$_3$] vs. t; 50 °C; mixing effects; pH=3.6; H$_2$CO$_3$

153

O_3-decomposition

DI/O_3 decomposition at RT - Effects of recirculation

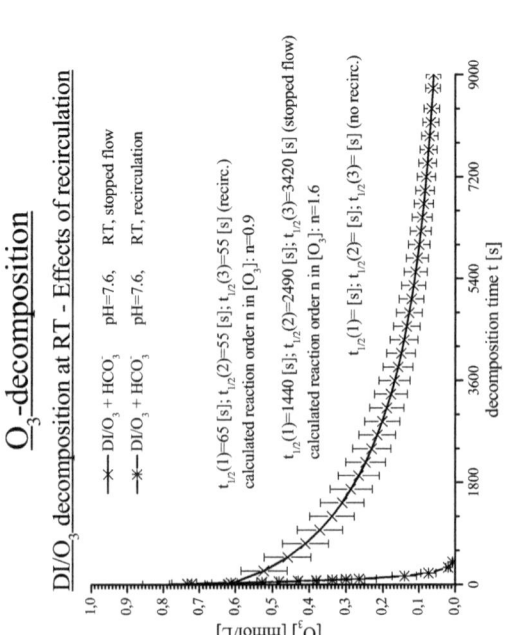

Figure 123 - [O_3] vs. t; RT; mixing effects; pH=7.6; HCO_3^-

O$_3$-decomposition

DI/O$_3$ decomposition at RT - Effects of recirculation

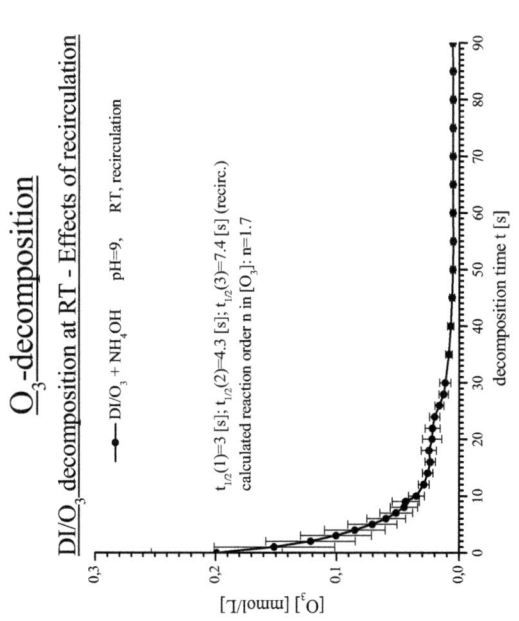

Figure 124 - [O$_3$] vs. t; RT; mixing effects; pH=9; NH$_4$OH

Appendix - Results

9.3. Resist characterization with IR spectroscopy

IR-study of PHS-type DUV-resists
Comparison M91Y & UV26 resist

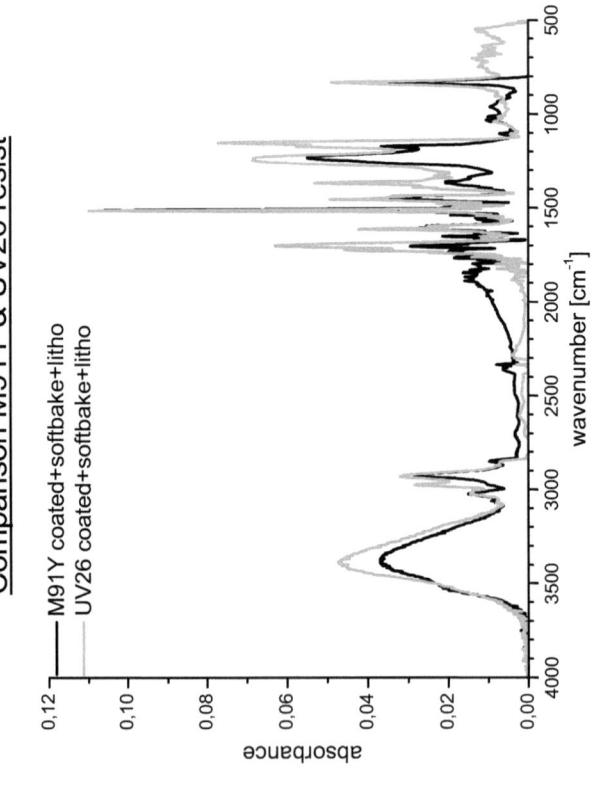

Figure 125 - Comparison of IR spectra M91Y; UV26 after exposure

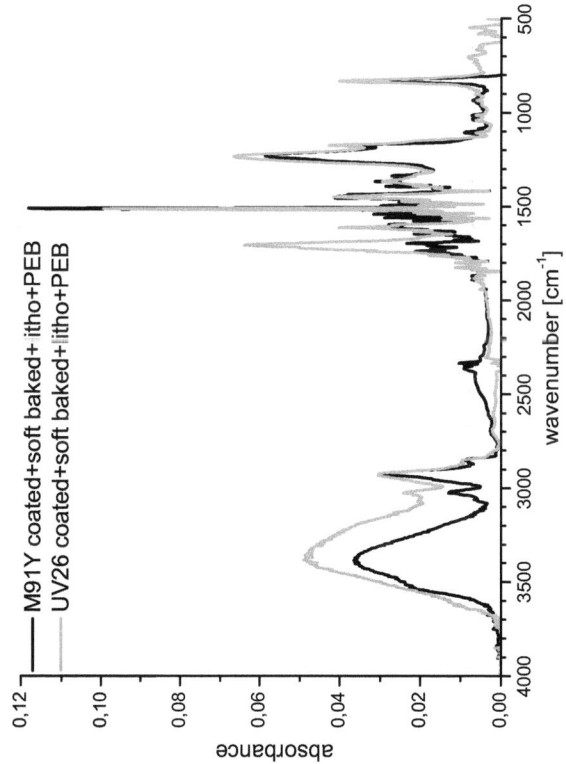

Figure 126 - Comparison of IR spectra M91Y; UV26 after PEB

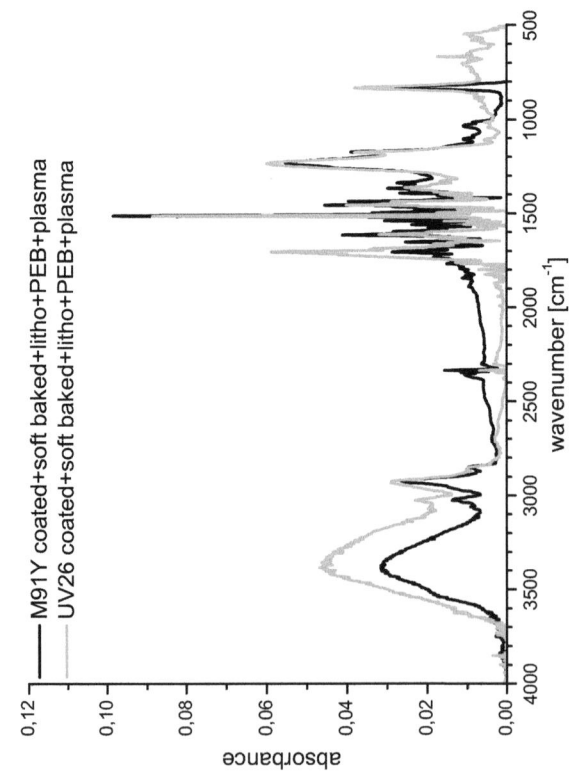

Figure 127 - Comparison of IR spectra M91Y; UV26 after plasma

Appendix - Results

9.5. Resist stripping

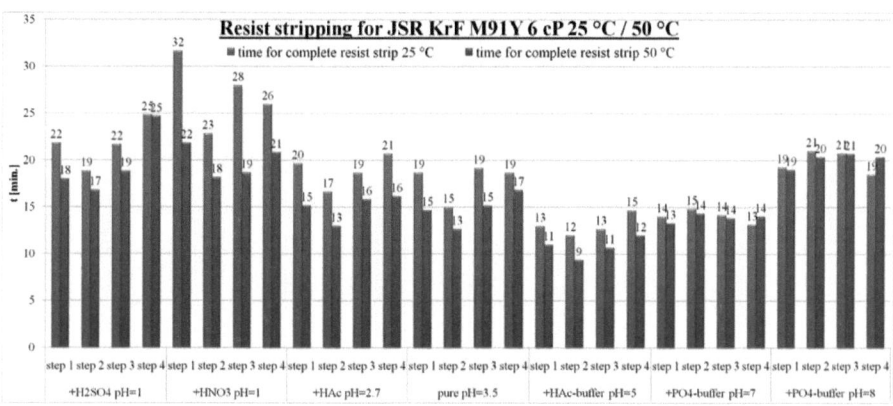

Figure 128 - Stripping efficiency comparison M91Y 25 °C vs. 50 ° (by pH)

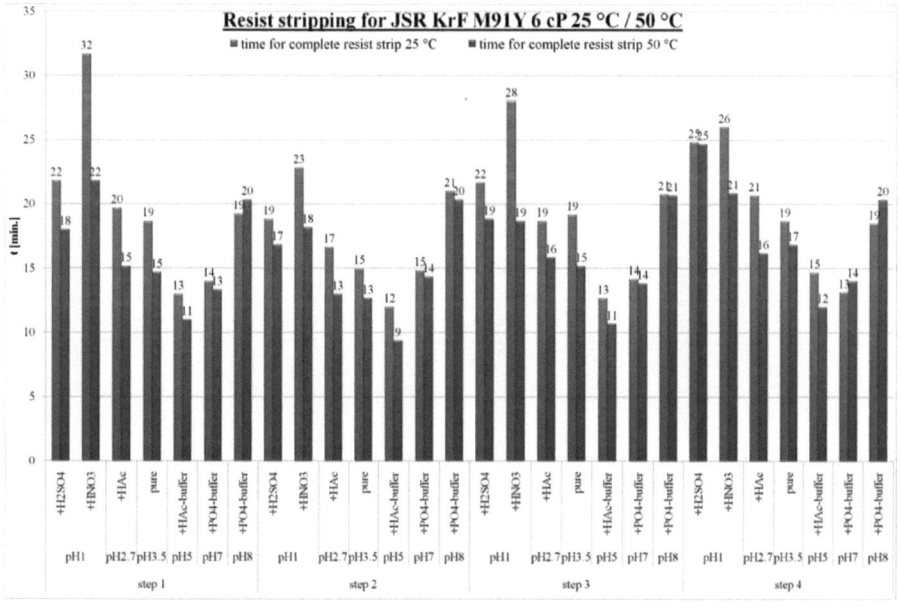

Figure 129 - Stripping efficiency comparison M91Y 25 °C vs. 50 ° (by step)

Appendix - Results

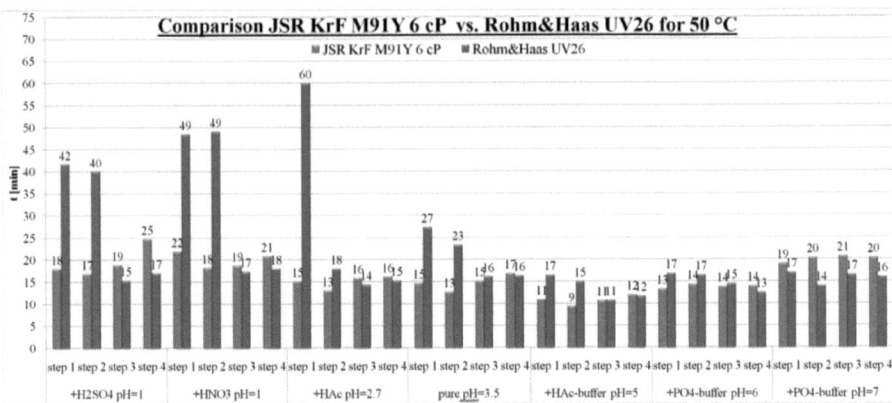

Figure 130 - Stripping efficiency comparison M91Y vs. UV26 at 50 ° (by pH)

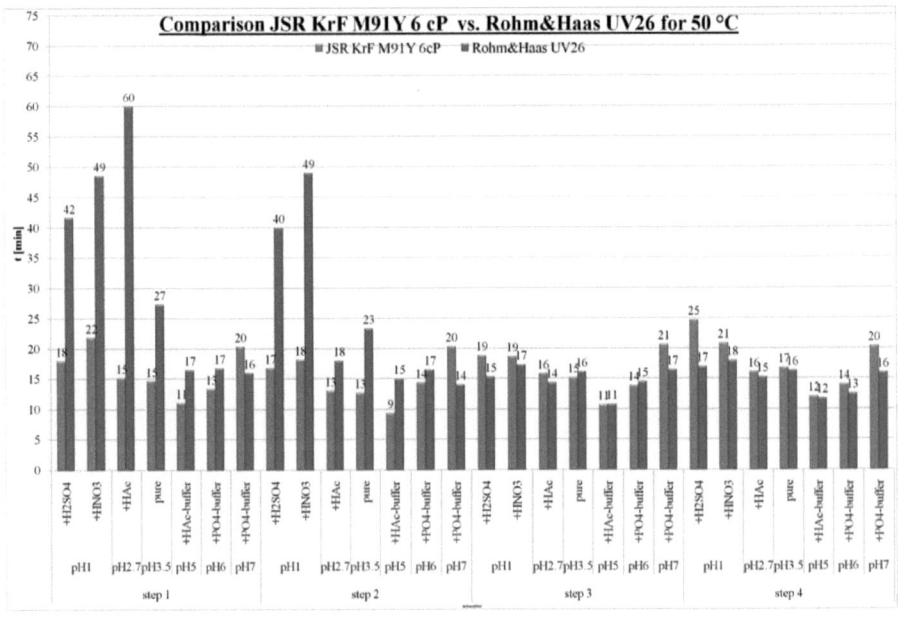

Figure 131 - Stripping efficiency comparison M91Y vs. UV26 at 50 ° (by step)

Appendix - Results

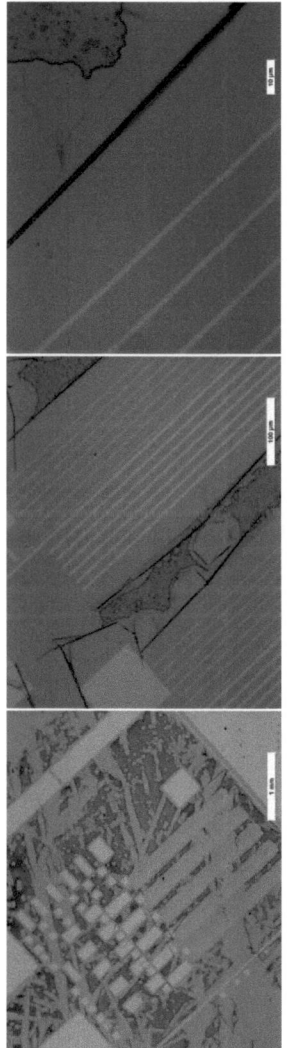

Figure 132 - (As 10^{16} cm^{-2}; 40 keV) - DI/O$_3$+buffer pH=5; RT; 1 h; UV; continues DI/O$_3$ flow

Figure 133 - (As 10^{16} cm^{-2}; 40 keV) - DI/O$_3$+buffer pH=6; RT; 1 h; UV; continues DI/O$_3$ flow

Appendix - Results

Table XIX - Figure comparison of 1 h stripping pH=12-13.5 with continues flow

		pure additive	additive + UV	additive + UV + O_3
pH = 12	KOH			
pH = 12	NH$_4$OH			

Appendix - Results

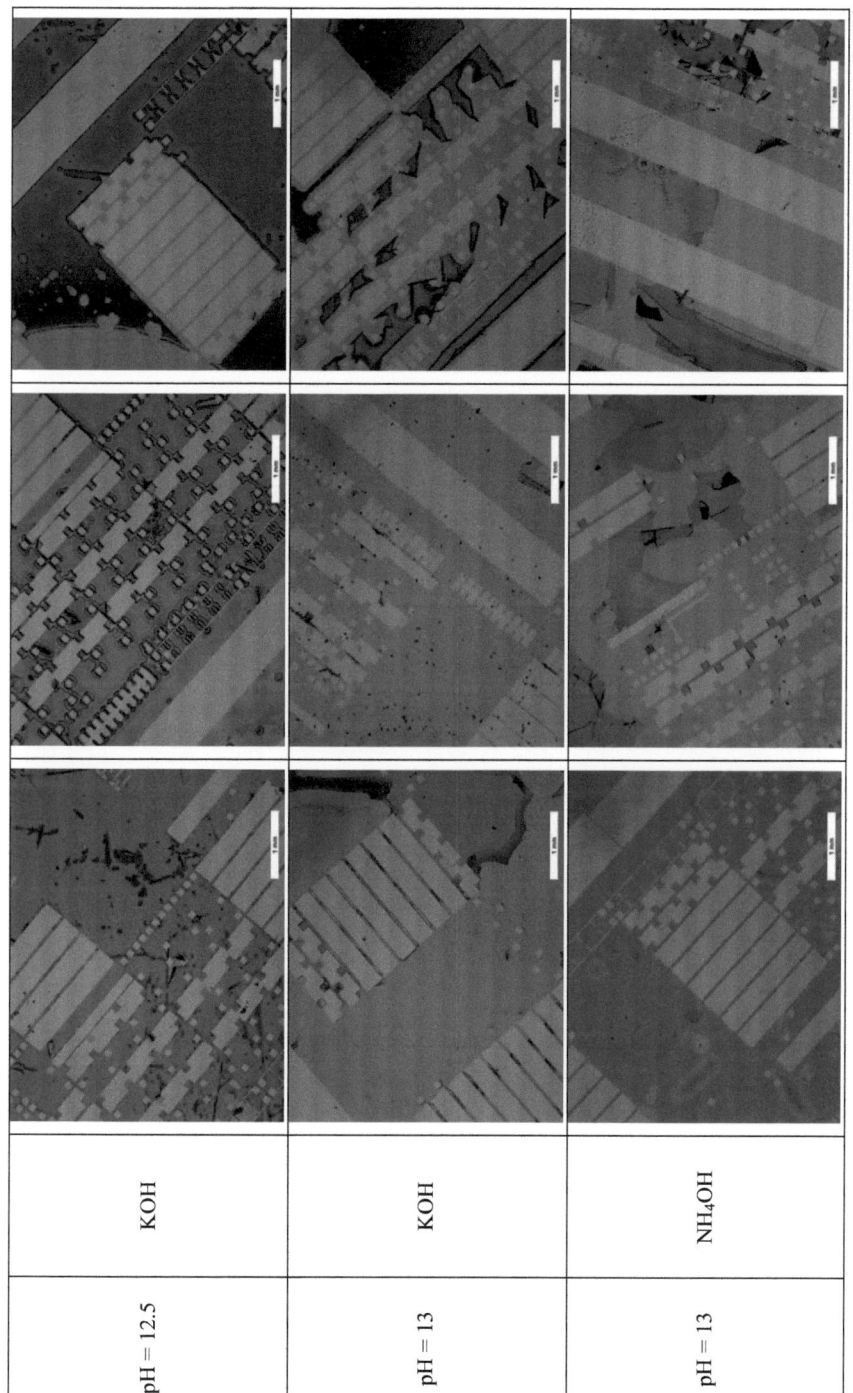

163

Appendix - Results

pH = 13	previous hard bake 130 °C 10 s			
pH = 13.12	pyrrolidin			
pH = 13.5	KOH			

164

Die VDM Verlagsservicegesellschaft sucht für wissenschaftliche Verlage abgeschlossene und herausragende

Dissertationen, Habilitationen, Diplomarbeiten, Master Theses, Magisterarbeiten usw.

für die kostenlose Publikation als Fachbuch.

Sie verfügen über eine Arbeit, die hohen inhaltlichen und formalen Ansprüchen genügt, und haben Interesse an einer honorarvergüteten Publikation?

Dann senden Sie bitte erste Informationen über sich und Ihre Arbeit per Email an *info@vdm-vsg.de*.

Sie erhalten kurzfristig unser Feedback!

VDM Verlagsservicegesellschaft mbH
Dudweiler Landstr. 99
D - 66123 Saarbrücken
www.vdm-vsg.de

Telefon +49 681 3720 174
Fax +49 681 3720 1749

Die VDM Verlagsservicegesellschaft mbH vertritt

Printed by Books on Demand GmbH, Norderstedt / Germany